Edward Heron-Allen

A Manual of Cheirosophy

Being a complete practical handbook of the twin sciences of cheirognomy and

cheiromancy, by means whereof the past, the present, and the future may be read

in the formations of the hands, preceded by an introductory argument

Edward Heron-Allen

A Manual of Cheirosophy

Being a complete practical handbook of the twin sciences of cheirognomy and cheiromancy, by means whereof the past, the present, and the future may be read in the formations of the hands, preceded by an introductory argument

ISBN/EAN: 9783337075453

Printed in Europe, USA, Canada, Australia, Japan

Cover: Foto ©berggeist007 / pixelio.de

More available books at **www.hansebooks.com**

A MANUAL

OF

CHEIROSOPHY

BEING

A Complete Practical Handbook

OF

THE TWIN SCIENCES

OF

CHEIROGNOMY AND CHEIROMANCY

BY MEANS WHEREOF

THE PAST, THE PRESENT, AND THE FUTURE

MAY BE READ IN THE FORMATIONS OF

THE HANDS

PRECEDED BY

An Introductory Argument upon the Science of Cheirosophy
and its claims to rank as a Physical Science

BY

ED. HERON-ALLEN

*Author of "Codex Chiromantiæ," "Dactylomancy," "A Discourse of Chyromancie
and of Mesmerism," "A Lyttel Boke of Chyromance," etc.
Joint Author of "Chiromancy; or, the Science of Palmistry," etc.*

WITH FULL-PAGE AND OTHER ILLUSTRATIONS BY
ROSAMUND BRUNEL HORSLEY

TENTH EDITION

LONDON:

WARD, LOCK & CO., LIMITED.

NEW YORK AND MELBOURNE.

PRINTED BY
HAZELL, WATSON, AND VINEY, LD.,
LONDON AND AYLESBURY.

Dedication.

If any good be in this Book of mine,
Or aught of Truth shine out therefrom, for Thee—
For Thine eyes' comfort and Thy soul's delight;
Oh Thou, where'er Thou art! it has been wrought.

And if approval be man's best reward,
Be Thine the meed; as is most justly due
To Her alone who called the writing forth:
But if, as may be, censure strike the ear,
Be his the blame who failed unworthily,
Blind with the precious prospect of Thy praise.

Nathless the Scribe shall guerdon win, and this:
The moment when Thy starry eyes, oh Thou,
Where'er Thou art! shalt read and understand
That all his labour has been but for Thee.

                                        E. H.-A

12. vii. 1885.

Gentile morbida, leggiadra mano
Cui fer le proprie mani d'amore;
Più dell'avorio candida e tersa;
Sparsa di varie pozzette molli;
Le cui flessibili lunghette dita
Dolce assottigliano in unghie vaghe
Arcate lucide rubicondette;
Distesa appressati al palpitante
Cor mio che cenere farsi già sento;
Potra resistere del caro sguardo
Allora a i fervidi raggi, onde fiamme
Soavi scendono ma troppo ardenti.

Paolo Rolli.

PREFACE.

I HAVE so far elaborated the Introductory Argument which precedes this work as to render any further preface unnecessary, and will, therefore, make use of this opportunity only by asking the reader carefully to peruse that Introductory Argument, wherein he will find expressed the object of this book, and, if such be necessary, its apology.

I have only to say that the following pages represent, in a condensed form, the studies and the personal observations of some years. Since the appearance of my former book on the science of Cheiromancy, many similar works have taken their place in the literature of this country; of these those resulting from the labours of Miss Rosa Baughan are the only volumes worthy of any serious consideration. I can only hope that by the perusal of the follow ing pages those students who have taken any interest in this Great Science will be afforded an opportunity of making deeper investigations into the ultimate, as well as into the proximate causes of the science than they have yet been able to make by the perusal of the hitherto standard works upon this subject.

ED. HERON-ALLEN.

St. John's, Putney Hill, S.W.
20th July, 1885.

TABLE OF CONTENTS.

	PAGES
PREFACE	7
INTRODUCTORY ARGUMENT.	17
SECTION I.—CHEIROGNOMY; OR, THE SHAPES OF THE HANDS	97
SUB-SECTION I.—CONCERNING THE HAND IN GENERAL AND THE INDICATIONS AFFORDED BY THE ASPECTS AND CONDITIONS OF ITS VARIOUS PARTS IN PARTICULAR	100
§ 1. The Palm of the Hand	100
§ 2. The Joints of the Fingers	102
§ 3. The Comparative Length of the Fingers	106
§ 4. The Fingers Generally	108
§ 5. The Finger Tips	111
§ 6. The Hairiness of the Hands	115
§ 7. The Colour of the Hands	115
§ 8. The Thumb	116
§ 9. The Consistency of the Hands	121
§ 10. The Cheirognomy of the Individual Fingers	125
§ 11. The Habitual Actions and Natural Positions of the Hands	129
§ 12. The Finger Nails	130

CONTENTS.

PAGES

SUB-SECTION II.—THE SEVEN TYPES OF HANDS, AND THEIR SEVERAL CHARACTERISTICS . . . 136

§ 1. The Elementary or Necessary Hand . . 139
§ 2. The Spatulate or Active Hand 140
§ 3. The Conical or Artistic Hand . . . 148
§ 4. The Square or Useful Hand . . . 155
§ 5. The Knotty or Philosophic Hand . . 159
§ 6. The Pointed or Psychic Hand . . . 164
§ 7. The Mixed Hand 170

SUB-SECTION III.—THE CHEIROGNOMY OF THE FEMALE HAND 174

SECTION II.—CHEIROMANCY; OR, THE DEVELOPMENTS AND LINES OF THE PALM . 183

SUB-SECTION I.—AN EXPLANATION OF THE MAP OF THE HAND 186

SUB-SECTION II.—GENERAL PRINCIPLES TO BE BORNE IN MIND 193

§ 1. As to the Mounts 193
§ 2. As to the Lines 196

SUB-SECTION III.—THE MOUNTS OF THE HANDS . 204

§ 1. The Mount of Jupiter 204
§ 2. The Mount of Saturn 206
§ 3. The Mount of Apollo 211
§ 4. The Mount of Mercury 214
§ 5. The Mount of Mars 217
§ 6. The Mount of the Moon 220
§ 7. The Mount of Venus 224

CONTENTS

	PAGES
SUB-SECTION IV.—THE LINES OF THE HAND	229
§ 1. The Line of Life	229
§§ 1. The Line of Mars	242
§ 2. The Line of Heart	242
§ 3. The Line of Head	247
§ 4. The Line of Saturn or Fortune	256
§ 5. The Line of Apollo or Brilliancy	261
§ 6. The Line of Liver or Health	266
§§ 1. The Cephalic Line, or Via Lasciva	268
§ 7. The Girdle of Venus	268
SUB-SECTION V.—THE SIGNS IN THE PALM	271
§ 1. The Star	271
§ 2. The Square	273
§ 3. The Spot	274
§ 4. The Circle	274
§ 5. The Island	274
§ 6. The Triangle	275
§ 7. The Cross	276
§§ 1. The "Croix Mystique"	277
§ 8. The Grille	277
§ 9. The Signs of the Planets	278
SUB-SECTION VI.—THE SIGNS UPON THE FINGERS	280
§ 1. Signs upon the First Finger or Index	281
§ 2. Signs upon the Second or Middle Finger	282
§ 3. Signs upon the Third or Ring Finger	282
§ 4. Signs upon the Fourth or Little Finger	283
§ 5. Signs upon the Thumb	283

CONTENTS.

	PAGES
SUB-SECTION VII.—THE TRIANGLE, THE QUADRANGLE, AND THE RASCETTE	285
§ 1. The Triangle	285
§§ 1. The Upper Angle	287
§§ 2. The Inner Angle	287
§§ 3. The Lower Angle	287
§ 2. The Quadrangle	288
§ 3. The Rascette and Restreintes	289
SUB-SECTION VIII.—CHANCE LINES	293
SUB-SECTION IX.—A FEW ILLUSTRATIVE TYPES	301
SUB-SECTION X.—MODUS OPERANDI	306
INDEX	313

LIST OF ILLUSTRATIONS.

Frontispiece. JOHANN HARTLIEB, AUTHOR OF "DIE KUNST CIROMANTIA."

PLATE	PAGES
I.—THE ELEMENTARY HAND	138
II.—THE SPATULATE OR ACTIVE HAND	141
III.—THE CONIC OR ARTISTIC HAND	149
IV.—THE SQUARE OR USEFUL HAND	154
V.—THE KNOTTY OR PHILOSOPHIC HAND	161
VI.—THE POINTED OR PSYCHIC HAND	165
VII.—THE MAP OF THE HAND	187
VIII.—CONDITIONS OF THE LINES	198

    Fig. 1. Spots upon a line.
    ,, 2. Sister lines.
    ,, 3. Forked terminations.
    ,, 5. Ascending and descending branches.
    ,, 6. Chained lines.
    ,, 7. Wavy lines.
    ,, 8. Broken lines.
    ,, 9. Capillaried lines.

IX.—SIGNS FOUND IN THE HAND . . . 202

    Fig. 10. The Star.
    ,, 11. The Square.
    ,, 12. The Spot.
    ,, 13. The Circle.
    ,, 14. The Island.
    ,, 15. The Triangle.
    ,, 16. The Cross.
    ,, 17. The Grille

ILLUSTRATIONS.

	PAGES
X.—Lines upon the Mounts of the Palm	209
XI.—Ages upon the Lines of Life and of Fortune	228
XII.—XVI.—Modifications of the Principal Lines 232, 235, 239, 243,	253
XVII.—Ditto. The Quadrangle and the Triangle	264
XVIII., XIX., and XX.—Chance Lines . 292, 295,	299

Head-pieces.—The Signs of the Zodiac and of the Seven Planets of Cheiromancy, in Silhouette Designs.

Rosamund Brunel Horsley, inbt. et delt.

An Introductory Argument
upon the Science of Cheirosophy and its Claims to rank as a Physical Science.

1. Sir Charles Bell—"The Hand, its Mechanism and Vital Endowments, as evincing Design and illustrating the Power, Wisdom, and Goodness of God."—*Bridgewater Treatise* (London, 1832).
2. Lud. Jul. Ern. de Naurath — "De Manuum morphologiâ et physiologiâ." (Berolini, 1833.)
3. John Kidd—"On the Adaptation of External Nature to the Physical Condition of Man."—*Bridgewater Treatise* (London, 1834).
4. G. C. Carus—"Ueber Grund und Bedeutung der verschiedenen Formen der Hand in verschiedenen Personen." (Stuttgart, 1846.)
5. Sir Richard Owen—"On the Nature of Limbs. A discourse delivered on Friday, Feb. 9th [1849], at an evening meeting of the Royal Institution of Great Britain." (London, 1849).
6. G. M. Humphry—"Observations on the Limbs of Vertebrate Animals, the Plan of their Construction, their Homology and the Comparison of the Fore and Hind Limbs." (Cambridge and London, 1860.)
7. G. M. Humphry—"On the Human Foot and Human Hand." (Cambridge and London, 1861.)
8. Arthur Kollmann—"Der Tast-Apparat der Hand der menschlichen Rassen und der Affen in seiner Entwickelung und Gliederung." (Hamburg and Leipzig, 1883.)

AN INTRODUCTORY ARGUMENT UPON THE SCIENCE OF CHEIROSOPHY AND ITS CLAIMS TO RANK AS A PHYSICAL SCIENCE.

Ἀντὶ πολλῶν ἂν ὦ ἄνδρες Ἀθηναῖοι, χρημάτων ὑμᾶς ἑλέσθαι νομίζω, εἰ φανερὸν γένοιτο τὶ μέλλον συνοίσειν τῇ πόλει περὶ ὧν νυνὶ σκοπεῖτε.

ΔΗΜΟΣΘΕΝΟΥΣ ʼΟΛΥΝΘΙΑΚΟΣ, Αʹ.

¶ 1. The Study of Eclectic Science

"IF the study of Phrenology, of Cheirosophy, and of the sciences which have for their aims the discovery of the true characters and instincts of men by the developments and appearances of their bodies, is merely a frivolous amusement, if such a study ceases for a moment to be a serious one, or if it is merely a distraction for enthusiasts, for people whose love of the marvellous becomes an insatiable greed,—it is in every way damnable and to be discouraged, because it results infallibly in superstition and error. BUT if it is based upon truth, men cannot give themselves up to the study with too much energy, not only on account of the material advantages to be derived therefrom, but because it is an important factor in the considerations which lead to the education of our children, who alone represent the progress of the future." It is with these words that Adrien Desbarrolles commences the preface of his elementary work, "The Mysteries of the Hand" (Paris, 1859); and I quote them at the head of this Introductory Argument, as the sentiments conveyed by them are the key-notes, and, as it were, the corner-stones of the composition of this work, my aim in writing this MANUAL OF CHEIROSOPHY having been simply to place before the world a concise and clearly comprehensible epitome of the principia of a science which opens a new page of the great book of nature to the student who will diligently read it, which gives to youth the experience and the foresight of age, and which endows all men who will study it with that foresight which, under the name of intuitive faculty, is the cherished possession of so few, enunciating and solving the great problem of "Know Thyself."[1]

Scope of this work.

¶ 2.
Method of discussion.

I have not set about the task of laying this science before a critical world with a view to its recognition as an exact science without being well aware of the difficulties to be surmounted, the prejudices to be overcome, and the apparent anomalies to be explained and reconciled with the dicta of physiology; but I shall endeavour categorically to discuss every point of the argument, shirking nothing which may seem adverse to my object, and giving undue prominence to nothing which may seem specially favourable thereto. I desire rather to enter upon the discussion after the manner of an uninterested third party, whose only desire is the clearing-away of doubt, and the establishment of a new science, whose full development *must* become an enormous advantage to mankind.

¶ 3.
The Hand

Without continuing to announce what I am *going*

---

[1] "Connais-toi toi-même ! Belle et sage maxime, à laquelle il est plus aisé à la généralité des hommes d'applaudir que de se conformer."—D'ARPENTIGNY, "La Science de la Main" (Paris, 1865). 3rd edn.

AN INTRODUCTORY ARGUMENT. 19

to do, let me begin by the consideration of that member with which we are particularly concerned, of that complex piece of mechanism wherein we find the radical principles of the science whose bases we are occupied in firmly establishing, and to which we have [if I am right] to look for the history of our lives.

¶ 4. Hand-shakin

There is no part of the human body which is more significant in its actions, which is more characteristic in its formation, than the Hand. I take as an illustration the most elementary indication afforded by the hand, an indication, the instinctive observation of which renders every one, to a certain extent, a Cheiromant,—I allude to hand-shaking, an action in itself symbolical, having been adopted in old days for the purpose of showing that the hand contained no weapon, so that there should be no danger of treachery between the hand shakers.[2] Has not every one experienced the feeling of confidence and good fellowship expressed by a good, firm grasp of the hand? the feeling of repulsion and discomfort which comes over one when one is given what a recent essayist calls " a hand like a cold haddock," or the instinctive distrust which awakens in us at a peculiar or uncomfortably individualized method of shaking hands?

¶ 5. Use of the har

It needs, I think, but very little to recommend the dictum of the ancient philosophers, that to his hands man owes his superiority over all other animals. One recognises the secondary influences of the hand in the writing of books and in the construction of articles of every-day use and necessity, which we cannot think would be produced without hands, though, as has often been remarked, it is not beyond the bounds of possibility for many things which we are accustomed to see done by the hands, to be effected

[2] *All the Year Round*, vol. iii., N.S., 1870, p. 467.

without them:—"We have daily before us," says Sir Charles Bell, "proofs of ingenuity in the arts, not only surviving the loss of the hand, but excited and exercised where the hands were wanting from birth. What is more surprising than to see the feet, under such circumstances, becoming substitutes for the hands, and working minute and curious things?"[1] This is, of course, very true; but when it occurs we are accustomed to look upon it more as a curiosity and a phenomenon, than as a natural consequence of the loss of this all-important member.

Loss of the hand.

¶ 6. The wants of man are greater and more varied than those of any other animal; and therefore, says Galen, he has had given to him what he alone of animals possesses, and what to him alone is necessary, viz. the hand. "For," says he, "some animals are bold and fierce, others are timid and gentle; some are gregarious and co-operate for their mutual sustenance and defence; others are solitary and avoid the society of their fellows; but all have a form or body accommodated to their natural dispositions and habits. Thus, the lion has powerful fangs and claws; the hare has swiftness of foot, but is otherwise defenceless. And the fitness of this arrangement is obvious;

The hand adapted to the wants of man. Galen.

A criminal without hands.

[1] The same author continues:—"Unfortunately, too, the most diabolical passions will be developed in some natures, and crimes committed which we might have supposed impossible from the power of execution being denied. The most remarkable instance of that was in a man who, from birth, was deprived of arms; as if possessed by a devil, this wretch had committed many murders before being discovered and executed. He was a beggar, who took his stand on the highway some miles from Moscow, on the skirts of a wood; his manner was to throw his head against the stomach of the person who was in the act of giving him charity, and, having stunned him, to seize him with his teeth, and so drag him into the wood."—SIR CHARLES BELL, "The Hand, its Mechanism and Vital Endowments, as evincing design and illustrating the power, wisdom, and goodness of God" (*Bridgewater Treatise:* London, 1832).

for those weapons with which the lion is furnished are as appropriate to his nature as they would be inappropriate to the timid hare, whose safety, depending entirely on flight, requires that swiftness of foot for which she is so remarkable. But to man, the only animal that partakes of divine intelligence, the Creator has given, in lieu of every other natural weapon or organ of defence, that instrument, *the hand*,—an instrument applicable to every art and occasion as well of peace as of war. Man, therefore, wants not a hoof, or horn, or any other natural weapon, inasmuch as he is able with his hand to grasp a much more effective weapon, the sword or spear; for a sword or a javelin are better arms than the nails, and cut and pierce more readily. Nor does he want strong nails like those of a horse, for a stick or a stone hits harder and better than such a nail. Besides which, natural weapons can be employed only in close conflict, while some of the weapons employed by man, such as javelins or arrows, are even more effective at a distance. And again, though man may be inferior to the lion in swiftness, yet, by his dexterity and skill, he breaks in to his use a still swifter animal, the horse, mounted on whose back he can escape from, or pursue the lion, or attack him at every advantage. He is enabled, moreover, by means of this instrument, to clothe himself with armour of various kinds, or to entrench himself within camps or fenced cities, whereas, were his hands encumbered with any natural armour, he would be unable to employ them for the fabrication of those instruments and means which give him such a decided advantage over all the other animals of creation."[4]

[4] The above is Dr. Kidd's translation (*vide* "On the Adaptation of External Nature to the Physical Condition of Man": London, 1833), of the opening chapter of Galen's work entitled, "Claudii Galeni Pergameni secundum Hippocratem medicorum principis,

¶ 7.
Manufactures

Galen goes on to point out that it is with the hand that we weave the garments which protect us from heat and cold in summer and winter, and construct the nets and snares with which we subjugate the brute creation. With the hand we fashion all the implements of art and science, as well as the altars and shrines of the immortal gods; "and, lastly, by means of the hand, man bequeaths to posterity in writing the intellectual treasures of his own divine imagination; and *hence* we who are living at this day are enabled to hold converse with Plato and Aristotle and all the

opus de usu partium corporis humani, magna cura ad exemplaris Græci veritatem castigatum universo hominum generi apprime necessarium Nicolao Regio Calabro interprete" (Parisiis, ex officinâ Simonis Colinæi, 1528.—The passage runs in the original:—Quædam enim sunt audacia, quædam timida, alia agrestia, alia mansueta, alia velut civilia et popularia, alia velut solitaria. Omnibus vero aptum ac habile est corpus, animæ moribus, et facultatibus. Equo quidem fortibus ungulis et juba est ornatum instructumque (etenim velox et superbum et non ignavum est animal). Leoni autem, utpote animoso et audaci, dentibus et unguibus validum. Ita autem et tauro et porco. Illi enim cornua, huic autem exerti dentes (quos Græci χαυλιοδόντας nominant) arma sunt innata. Cervo autem et lepori (timida enim sunt animalia) velox quidem est corpus, nudum autem omnino et inerme. Timidis enim, opinor, velocitas, arma vero audacibus conveniebant. Neque igitur timidum aliquod armavit natura, neque audax et strenuum quodpiam nudum destituit. Homini autem (sapiens enim est hoc animal, et solum eorum, quæ sunt in terrâ, divinum) pro omnibus simul defensoriis armis, manus dedit, instrumentum ad omnes perinde artes accommodum et necessarium, pacique non minus quam bello idoneum. Non igitur indiguit cornu sibi innato, quum meliora cornibus arma in summis manibus, quandocunque volet, possit accipere. Etenim ensis et hasta majora sunt arma et ad incidendum promptiora, quam cornua. Sed necque indiguit ungula. Nam lapis et lignum quavis ungula quassant fortius violentiusque. Ad hæc, necque cornu, necque ungula, quidquam nisi cominus, agere possunt. Hominum vero arma, eminus juxta ac cominus agunt, telum videlicet ac sagitta magis quam cornua, lignum autem et lapis, magis quam ungula. Sed, velocior est homine leo. Quid hoc tandem est? Homo siquidem equum domuit sapientia et manibus, velocius leone animal, quo utens, et

venerable sages of antiquity."[5] Thus the hand keeps constantly before us the proofs of the special adaptations of the various parts of the body to the uses to which the parts are to be put. "In man," says Professor Owen [*vide* note [23], p. 36], "whilst one pair of limbs is expressly organized for locomotion and standing in the erect position, the other pair is left free to execute the manifold behests of his rational and inventive will, and is exquisitely organized for delicate touch and prehension, emphatically called 'manipulation.'"

¶ 8. *Why man alone has hands. Aristotle. Anaxagoras.*

Laying aside, therefore, with these references all consideration of the *works* of men's hands, let us turn to those authors who have laid down the axiom that man is the wisest of all animals, *not* because he has hands, but that he has hands *because* he is the wisest of all animals. It was Anaxagoras who remarked that, because man had hands, he was the wisest of all animals; but Aristotle corrected him by

subterfugit, et persequitur leonem, et sedens, ex alto humilem infernumque illum percutit. Non igitur est nudus, necque inermis necque vulnerari facilis, necque calciamentorum expers homo, sed ipsi est thorax ferreus quandocunque libet, omnibus coriis difficilius perforabile organum. Est autem et calciamentorum multimoda species, est et armorum, est et operimentorum. Non igitur thorax solum, sed et domus, et murus, et turris, sunt hominis operimenta munimentaque. Si autem innatum esset ei cornu in manibus, vel aliquod tale aliud armorum defensiorum, non utique posset uti manibus ad domus constructionem, vel muri, vel hastæ, vel thoracis, vel alicujus alterius similis.

[5] "His manibus homo vestem texuit et retem connexuit, et nassam, et sagenam, et velum. Quare non solum his quæ in terra animalibus, sed eis etiam quæ in mari et aere dominatur. Talia quidem homini ad fortitudinem arma sunt manus. Ut autem pacificum et politicum animal homo, manibus leges scripsit, et templa, et simulachra diis erexit, et navem, fistulam, lyram, scalpellum, forcipem, et alia universa artium instrumenta construxit, et commentarios etiam speculationis eorum, scriptos reliquit, tibique liceat literarum et manuum beneficio etiam nunc colloqui cum Platone, cum Aristotele, cum Hippocrate, et aliis veteribus."—*Ib.*

saying that it is because man is the wisest of all animals that he has been given hands; and in this view of the case Galen concurs.[6]

¶ 9. *The hand is the organ of the mind.*

The hand is essentially the organ of the mind, the medium of its expression, and the instrument whereby its promptings are carried into execution. "We first see the hand," says Sir Charles Bell, "ministering to man's necessities and sustaining the life of the individual; in a second stage of his progress, we see it adapted to the wants of society, when man becomes a labourer and an artificer; in a state still more advanced, science is brought in aid of mechanical ingenuity, and the elements which seem adverse to the progress of society become the means conducing to it. The seas, which at first set limits to nations, and grouped mankind into families, are now the means by which they are associated. Philosophical chemistry has associated the elements to man's use; and all tend to the final accomplishments of the great objects to which everything, from the beginning, has pointed—the multiplication and distribution of mankind, and the enlargement of the sources of man's comfort and enjoyment, the relief from too incessant toil, and the consequent improvement of the higher faculties of his nature." Is it unreasonable to pay particular attention to the instrument by which all these things are accomplished, and to regard it as something more than a mere member or organ of the body? Aristotle was quite right when he called the

Aristotle.

[6] ARISTOTLE, ΠΕΡΙ ΖΩΩΝ ΜΟΡΙΩΝ Βιβλ. Δ'., Κεφ. l.,—"'Ἀναξαγόρας μὲν οὖν φησὶ διὰ τὸ χεῖρας ἔχειν φρονιμώτατον εἶναι τῶν ζῴων ἄνθρωπον· εὖ λόγον δὲ διὰ τὸ φρονιμώτατον εἶναι χεῖρας λαμβάνειν." And Galen remarks on this passage (Op. cit., lib. I., c. I): "Ita quidem sapientissimum animalium est homo. Ita autem et manus sunt organa sapienti animali convenientia. Non enim quia manus habuit, propterea est sapientissimum, ut Anaxagoras dicebat; *sed quia sapientissimum erat,* propter hoc manus habuit, ut rectissime censuit Aristoteles."

hand the organ of the organs,[7] the active agent of the passive powers of the entire system; and Desbarrolles has followed in his footsteps, when he has said, that like as man is a condensation of the universe, a microcosm, so is the hand a condensation of the man.

Not only are "hands given us for our defence,"[8] but with weapons in our hands, and with our hands alone, we can measure and regulate the force and extent of our defensive and offensive actions; and it is by this power of regulation that we recognize the presence of what is known as a muscular sense. It is to this power of the regulation of force that Sir Charles Bell refers, after alluding to that magnificent passage in the Odyssey, where Ulysses deliberates upon the force of the blow he deals to the beggar Irus,[9] saying:—
"By such arguments, I have been in the habit of showing that we possess a muscular sense; and that, without a perception of the condition of the muscles previous to the exercise of the will, we *could not command them* in standing, far less in walking, leaping, or running. And as for the hand, it is not more the freedom of its action which constitutes its perfection, than the knowledge which we have of these motions, and our

¶ 10.
Offensive and defensive functions.

Regulation of force.

Homer.

Muscular sense.

---

[7] ΠΕΡΙ ΖΩΩΝ ΜΟΡΙΩΝ Βιβλ. Δ'., Κεφ. L.—"Ἡ δὲ χεὶρ ἔοικεν εἶναι οὐχ ἓν ὄργανον ἀλλὰ πολλά· ἔστι γὰρ ὡσπερεὶ ὄργανον πρὸ ὀργάνων."

[8] HOMER, ΙΛΙΑΣ, Βιβλ. Ν'., l. 813:—
"Ἡ θήν πού τοι θυμὸς ἐέλπεται ἐξαλαπάξειν,
νῆας· ἄφαρ δέ τε χεῖρες ἀμύνειν εἰσὶ καὶ ἡμῖν."

[9] HOMER, ΟΔΥΣΣΕΙΑΣ Βιβλ. Σ'., l. 90:—
"Δὴ τότε μερμήριξε πολύτλας δῖος Ὀδυσσεύς,
Ἢ ἐλάσει', ὥς μιν ψυχὴ λίποι αὖθι πεσόντα,
Ἠέ μιν ἦκ' ἐλάσειε, τανύσσειέν τ' ἐπὶ γαίῃ.
Ὧδε δέ οἱ φρονέοντι δοάσσατο κέρδιον εἶναι,
Ἦκ' ἐλάσαι, ἵνα μή μιν ἐπιφρασσαίατ' Ἀχαιοί.
Δὴ τότ' ἀνασχομένῳ, ὁ μὲν ἤλασε δεξιὸν ὦμον
Ἶρος, ὁ δ' αὐχέν' ἔλασσεν ὑπ' οὔατος, ὀστέα δ' εἴσω
Ἔθλασεν· αὐτίκα δ' ἦλθε κατὰ στόμα φοίνιον αἷμα·" etc.

consequent ability to direct it with the utmost precision." So it will be observed that among the lower animals the nearer approach to a hand that we find, the higher the grade of intelligence in the animal; and this has been noted by both Lucretius and Cicero, who point out the fact that the elephant has in its trunk the practical equivalent of the human hand.[10]

Lucretius. Cicero.

¶ 11. *Perfection of the hand.* One of the first points which obtrude themselves upon the student of anatomy is the absolute perfection of the human hand as regards its construction, and the uses to which it is adapted. In no other combination of bones, muscles, and nerves, and in no other animal do we find a perfection which results in such superiority with regard to strength, variety, extent, and rapidity of motion; and this perfection undoubtedly resulting from the intimate relations which exist between the hand and the intellect, we are irresistibly impelled to ask with Sir Charles Bell :— Is it nothing to have our minds awakened to the perception of the numerous proofs of design which present themselves in the study of the hand, to be brought to the conviction that everything in its structure is orderly and systematic, and that the most perfect mechanism, the most minute and curious apparatus, and sensibilities the most delicate and appropriate, are all combined in operation that we may move the hand? As Galen remarks: "Let us then scrutinize this member of the body and inquire, not simply whether it be in itself useful for all the purposes of life, and adapted to an animal endued with the highest intelligence, but whether its entire

Galen.

---

[10] LUCRETIUS, "De Rerum Naturâ," lib. ii., l. 536 :—

"Sicut quadripedum cum primis esse videmus
In genere *anguimanus elephantos*," etc ;

and CICERO, in his "De Naturâ Deorum," lib. ii., 123, says :—
"*manus etiam data elephanto est.*"

AN INTRODUCTORY ARGUMENT.   27

structure be not such, that it could not be improved by any conceivable alteration."[11]

¶ 12. *The hand as an indication of race.*

A writer in the *Anthropological Review* [vol. iii., 1865, p. 346], in a notice of Mr. R. Beamish's work "The Psychonomy of the Hand" (London, 1865), makes the following extremely pertinent remarks:—
"We have given this subject more attention than the work under consideration deserves, because we think that the hand has been hitherto unwisely neglected as an index of race. . . . It is very obvious that we have yet much to learn in this direction; we want more facts; we have not yet the data which would warrant even a plausible hypothesis. . . . Let us endeavour to discover if there be ethnic characteristics attaching to the extremities as well as to the cranium; let us first settle the great question of the racial hand, and then we shall be the better prepared to descend into the details of individual specialty." This writer had undoubtedly reason on his side, and I think that there is but very little ambiguity between the hands of various races at the present moment; we shall see, as we get further into the subject of Cheirognomy, how certain shapes of hands predominate among the English, the Germans, and the French; *English, German, and French hands.* and again, we shall notice the widely differing characteristics of the hands of meridional and septentrional, of oriental and occidental nations. Again, we shall see how different characters and mental calibres admire different shapes of hands, according to the characteristics which those shapes represent in Cheirognomy; and, if evidence of this were required, we should find it in the preference

---

[11] GALEN, Op. cit., lib. i., p. 4. "Agedum igitur, hanc ejus partem primam expendamus, non perscrutantes sit ne hæc plane simpliciterque utilis, aut an sapienti animali conveniens, sed num eam omnino constitutionem habeat, qua meliorem aliam habere non potuit."

Ancient and modern hands. which the ancients showed for large hands [exemplified in all their statuary [12]] as compared with the modern admiration for a small hand; we shall see that a large hand is always an indication of force and power as opposed to the more *spirituel* disposition denoted by the possession of small hands.

¶ 13.
Construction of the hand. Thus it will have been appreciated that the hand is the most perfectly constructed and constituted member of the body, that it is the member most typical of the sentient soul of man, and that it is an organ eminently fitted for the expression and development of the highest human faculties. It is not surprising, *Symbols.* therefore, that the symbols and symbolical actions in which we find the hand most prominent, are practically without number or limit; and that, in all times and in all countries, the hand has been accepted as the recognized embodiment of all force and intelligence.

¶ 14.
Symbolical expressions. The first point which will strike the student is the constant use of the word "hand" among the ancients to denote anything indicative of force or strength.[13] We find this in such phrases as χεὶρ σιδηρᾶ, a grappling iron; προσφέρειν χεῖρας, to apply force; χειρῶν ἄρχειν, to begin a fight; χεῖρας ὀρέξαι, to entreat; ἐν χερσί, a close fight; πρὸ χειρῶν, at hand; ἀπὸ χειρὸς, off hand; ὑπὸ χειρός, under the power; so also we find the words χείριος, subject to; χειρο-δίκης, assertion of right by force; χειρο-ήθης, submissive; χειρόω, I master or subjugate; χείρωμα, a conquest; and a number of other Greek works into which the word χείρ is introduced with the meaning of *force* or *power*.

[12] It is probably in deference to this obvious preference that the restorer of the Apollo Belvidere gave to that unfortunate statue the dreadful hand which makes one shudder amid the beauties of the Octagon Court at the Vatican.

[13] Witness also our modern expressions, "the hand of God" or "the finger of God," of the use of which, and of this use of the word "hand," there is no better example than the title of a compara-

AN INTRODUCTORY ARGUMENT. 29

Analogous to these is the old English expression with regard to size, height, or strength, such as we find it in passages of Shakespeare, as for instance, "He is a tall man *of his hands*" [*Merry Wives of Windsor*, I., 4.], or, as referring to strength :—"I'll swear to the Prince that thou art a tall fellow *of thy hands.*" "I know thou art a tall fellow *of thy hands*" [*Winter's Tale*, V., 3.], and "I am a proper fellow of my hands" [*Henry IVth.*, II.], and so on *ad libitum*. ¶ 15. Shakespeare's use of the word to denote force.

The symbols and customs connected with the hand are as interesting as they are numerous, many of them being of religious import; indeed, the most universal custom of folding the hands or of raising them in prayer being of the remotest antiquity. Aristotle refers to it,[14] and it undoubtedly has its origin in the symbolism that by folding the hands all power is surrendered by the person praying. It is for this reason that in Persia and in other Oriental nations it is customary to fold or hide the hands in the presence of a superior, thus symbolizing an abrogation of the will. ¶ 16. Folding of the hands in prayer.

Aristotle.

Oriental customs.

The giving of the hand has always been a token of peace and submission [whence arises, as I have mentioned above, our custom of shaking hands]; thus the bride gives her hand in the marriage ceremony in token of her submission to her husband, and in the middle ages part of the ceremony of the homage was the act of the feudatory placing his hands in those of the feudal lord or suzerain. ¶ 17. Giving of the hand.

In like manner we kiss the hands of princes in token of our submission to them, and we kiss the hands of fair women in acknowledgment of our ¶ 18. Kissing hands.

---

tively recently published work: "La Main de l'Homme et le Doigt de Dieu dans les malheurs de la France," par J. C., Ex-aumonier dans l'armée auxiliaire (Paris, 1871).

[14] ARISTOTLE, ΠΕΡΙ ΚΟΣΜΟΥ, Κεφ. Ζ':—Καὶ γὰρ πάντες οἱ ἄνθρωποι ἀνατείνομεν τὰς χεῖρας εἰς τὸν οὐρανὸν εὐχὰς ποιούμενοι."

Homer. allegiance. We all remember the exquisitely touching account of how Priam in his pathetic appeal humbled himself to kiss the hands of Achilles, the murderer of his children, when begging him to restore the body of Hector.[15] In the earlier days of the church it was customary to kiss the hands of the bishops, *Pliny.* and later Pliny tells us that Cæsar only allowed his hands to be kissed by persons of the higher ranks, the common people kissing their own hands on coming into his presence, as they did on entering their temples.

¶ 19. Kissing hands for salutation.
We have also the custom of saluting people by kissing our own hands to them, concerning which Brand[16] makes the following very interesting observation:—" I have somewhere read that the custom of kissing hands by way of salutation is derived from the manner in which the ancient Persians worshipped the sun, which was by first laying their hands upon their mouths, and then lifting them by way of adoration. A practice which receives illustration from a passage in the Book of Job,[17] a work replete with allusions to ancient manners:—'If I beheld the sun when it shined, or the moon walking in brightness; and my heart hath been secretly enticed, or my mouth hath kissed my hand.'"

¶ 20. The number "5."
In Morocco the number "Five" is never mentioned in the presence of the emperor, as it suggests the

---

[15] HOMER, Iliad, book xxiv., l. 477:—
" Τοὺς δ' ἔλαθ' εἰσελθὼν Πρίαμος μέγας, ἄγχι δ' ἄραστας,
Χερσὶν 'Αχιλλῆος λάβε γούνατα, καὶ κύσε χεῖρας,
Δεινὰς, ἀνδροφόνους, αἳ οἱ πολέας κτάνον υἷας."

l. 504:—
"'Εγὼ δ' ἐλεεινότερός περ,
Ἔτλην δ', οἷ οὔπω τις ἐπιχθονίοις βροτὸς ἄλλος,
'Ανδρὸς παιδοφόνοιο ποτὶ στόμα χεῖρ' ὀρέγεσθαι."

[16] JOHN BRAND, "Popular Antiquities of Great Britain." Edited by W. Carew Hazlitt. (London, 1870.)

[17] Job xxxi. 26-7.

hand which [as the holder of the sceptre] is there the emblem of all power; and the Turks consider the hand to represent the Deity, the fourteen joints being to them, as it were, the beads of a rosary, and the five fingers each representing a pious precept: viz., "Belief that there is but one Allah, and that Mahammud is his prophet," "The necessity of prayer," "The righteousness of Almsgiving," "Observation of the Radaman," and "The Journey to Mecca." *The Turkish rosary.*

The customs of raising the hand in voting or in taking an oath are as ancient as that of raising the hands in prayer. The Athenian word for to vote [in the ἐκκλησία or elsewhere] was χειροτονέω, the word signifying a vote being χειροτονία; and up till a comparatively late period, in our courts of law the prisoner, in pleading to an indictment, or a witness taking the oath, used to raise the right hand; indeed, this custom still holds good in Scotland. To swear "by the hand" used to be a common enough form of oath, and we find it continually repeated in one form or another in the works of Shakespeare, as, for instance, in such passages as, "By these *ten bones*, my lords, he did speak" (2 *Henry VI.*, Act I., sc. 3); "So do I still by these *pickers and stealers*" (*Hamlet*, Act III., sc. 2); "By this *hand*, I will supplant some of your teeth" (*Tempest*, Act III., sc. 2); and again, "By this *hand*, they are scoundrels and subtractors" (*Twelfth Night*, Act I., sc. 3); and the same formula occurs continually in the works of all the older dramatists. ¶ 21. *Voting and swearing with the hands.* Oaths. Shakespeare.

The ancients considered it to be the most terrible punishment that they could inflict upon their enemies to cut off one or both of their hands; and Xenophon and other authors tell us that after the battle of Ægos Potami, Lysander condemned all the Athenian prisoners to death for having decreed that, should they win, all the prisoners they took should have ¶ 22. *Ancient punishment of cutting off hands. Xenophon.*

their right hands cut off, Adimantos only being spared, because he had opposed this horrible decree in the Assembly;[18] and we find in the "De Bello Gallico" that, on one occasion, Cæsar had the hands of all his Gaulish prisoners cut off, as an example to the other tribes. In like manner, after the siege of Numantia, Scipio Africanus is said to have punished, by the amputation of their hands, four hundred of the inhabitants of the city of Lutea, for the assistance they had rendered the besieged.

Cæsar.

Scipio Africanus.

¶ 22. *Religious symbols. Blessings.*

I have adverted above to the religious symbolisms connected with the hand; the priest's blessing being given with the whole hand, and the episcopal blessing being given with the thumb and first two fingers only extended; and the reason of this is that these three represent the Trinity. The thumb is the representative of Unity in the Godhead; and in old books on the Ritual we find it laid down that in baptism the cross is to be traced on the infant's forehead with the thumb (*pollice*). The first finger is the emblem of Christ as the Indicator of God's will, the great Revealer and Declarer of God's will to mankind, the only finger that can stand upright of itself alone. [So, too, with the heathen the first finger was taken to be the representative of the god Jupiter, and this is also adhered to (as we shall presently see) in Cheiromancy.] The second finger represents ecclesiastically the Holy Ghost, and in this manner the three fingers thus extended represent the Trinity in the episcopal blessing. We find this again in the marriage service in the placing of the ring [which *used* to be placed upon the thumb] on the third finger, the old practice being to place the ring first on the thumb,

The emblem of the Trinity.

The first finger.

The second finger.

Marriage service.

[18] XENOPHON, ἙΛΛΗΝΙΚΟΝ, Βιβλ. Β´., Κεφ. ά (31):—"Ἐνταῦθα δὴ κατηγορίαι ἐγίγνοντο πολλαὶ τῶν Ἀθηναίων ἅ τε ἤδη παρανενομήκεσαν καὶ ἃ ἐψηφίσμενοι ἦσαν ποιεῖν, εἰ κρατήσειαν τῇ ναυμαχίᾳ, τὴν δεξίαν χεῖρα ἀποκόπτειν τῶν ζωγρηθέντων πάντων."

then on the first finger, then on the second finger, and, lastly, to leave it on the third finger, in token that after his allegiance to God in the Trinity, man's whole and eternal allegiance is given to his wife; the *ring* being the symbol of eternity.

¶ 24. Hand-charms.
The thumb and two fingers thus extended are familiar to all of us from the little coral charms fashioned on this wise which are worn by the Neapolitans as a talisman against the *jettatura* or evil eye; and I purchased, a short while since in Rome, a small silver hand whose fingers were thus arranged, and on various parts of which were stamped in relief various symbols, amongst which were a bust of Serapis, a knife, a serpent, a newt, a toad, a pair of scales, a tortoise, and a woman with a child. This talisman is known as the "Mano Pantea," and is an exact copy of one executed in bronze formerly in the Museo di Gian Pietro Bellori at Rome, and is said to be a most potent charm "contro del fascino." [19] As a matter of every day practice, the Neapolitan or the Sicilian averts the evil eye by means of what are known as the Devil's Horns, *i.e.*, the second and third fingers folded over the thumb, leaving the first and fourth fingers sticking out.

Mano Pantea.

The evil eye.

¶ 25. Raising of the hands.
The raising of the hands has always been recognized as a sign of peace and good faith, probably

[19] Questa mano esattamente imitata in piccola proporzione da quella di bronzo al naturale che già era nel Museo di Gian Pietro Bellori in Roma, ed ora non si sa dove sia, ma se ne ha il disegno nell' Opera del Grevio, vol. xii., page 763, donde fu ricavata. L'alto delle dita e i simboli che la ricoprono cioè il busto di Serapide, divinità propizia agli uomini, il coltello, il serpe, il ramarro, il rospo, la bilancia, la tartaruga, due vasi, la figura della donna col bambino, e un altro oggetto ignoto, formano un gruppo di simboli, che uniti insieme si credevano essere potenti a respingere gli effetti del fascino; e queste mani grandi lo tenevano in casa per proteggerla contro ogni cattivo influsso della magia o del mal occhio, quelle piccole le portavano indosso per esserne difesi.

Mano Pantea.

Xenophon.

from the fact that with the hands in the air the probabilities and possibilities of treachery or hostile demonstration, are minimised, and we have Xenophon's authority that this was also the case in the days when he lived.[20]

¶ 26.
The Thumb.
Gladiatorial
shows.

The ancient Romans used to indicate whether they desired the life or the death of the fallen gladiator by hiding the thumb (*premere pollicem*), or by turning it downwards (*vertere pollicem*); and in many other ways the importance of the thumb has been recognized in ancient customs. Ducange tells us that the ancient

Contracts

formula on the execution of documents used to be in the middle ages, "Witness my *thumb* and seal;" and even to-day, on the conclusion of a bargain in Ulster, the natives are wont to say, "We may lick *thooms* on that." To bite the thumb was in the middle ages

.Challenges.

equivalent to a challenge; thus we have it in the opening scene of *Romeo and Juliet*, where Sampson remarks, "I will bite my thumb at them; which is a disgrace to them if they bear it," to which Abram exclaims, "Do you bite your thumb at us, sir?" and the quarrel begins. In the classic ages cowards, who did not

"Poltroons."

want to go to the war, cut off their thumbs so as to render it impossible for them to handle a sword; and thus from the words *pollice truncatus* comes our word "poltroon," signifying coward.

¶ 27.
Superstitions.

There are many superstitions connected with the hand, such as that one whereby the hand of a hung man was said to cure warts and tumours if stroked over the affected spot; and a writer in *Fraser's Magazine* [vol. xxxvi., 1847, p. 293] tells of this remedy having been applied to the neck of a woman at the execution of Dr. Dodd in 1777. Hazlitt also, in his edition of Brand's "Antiquities" [*vide* note [16],

---

[20] ΧΕΝΟΡΗΟΝ, ΚΥΡΟΥ ΠΑΙΔΕΙΑΣ Βιβλ. Δ'., Κεφ. β'. (17):—
"Ἐκ τούτου πέμπει τὸν ἕτερον αὐτῶν πρὸς αὐτούς, προστάξας λέγειν· εἰ φίλοι εἰσίν, ὡς τάχιστα ὑπαντᾶν τὰς δεξιὰς ἀπατείναντας."

AN INTRODUCTORY ARGUMENT. 35

p. 30], gives a more precise account of the same thing occurring at the execution of the murderer Crowley, at Warwick, in the year 1845. Most of my readers will know the "Hand of Glory" from having read the Rev. R. H. Barham's "Ingoldsby Legend" of that name. The Hand of Glory was a talisman, apparently much used at one time for the commission of burglaries, for its properties were [we are told] that it would open closed doors, and when presented to people would deprive them of all power to move, though they might be awake; it was invisible to all except to him who held it, and prevented sleeping people from waking: as "Ingoldsby" hath it:— *Hand of Glory*   *Ingoldsby.*

> "Now open lock, to the Dead Man's knock!
> Fly bolt and bar and band!
> Nor move nor swerve, joint, muscle, or nerve,
> At the spell of the dead man's hand!
> Sleep all who sleep!—Wake all who wake!—
> But be as the dead for the Dead Man's sake!"

A full receipt [with an illustration] for the preparation of the Hand of Glory from the right hand of a man who has been hung in chains at a cross road, and for the composition of the candle to be held by it, may be found at page 104 of "Les Secrets du Petit Albert."[21]

Thus it will be seen from the above selected specimens that the symbolisms and superstitions connected with the hand are practically without number, and prove the effect which the uses and perfection of the hand produce in the most enlightened, as well as in the most ignorant minds. Let us, therefore, turn to the examination of the physical construction of this marvellous and perfectly constituted member.

¶ 28. Volume of symbolisms.

[21] "Secrets merveilleux de la magie naturelle et cabalistique du Petit Albert" (Lyons, 1776).

¶ 29.
Physiology of the hand.

It would be impossible, as it would be inexpedient and unnecessary, to embark in this place upon a long dissertation concerning the anatomy, physiology, and histology of the human hand; interesting as the subject is, I must refer my readers to the perusal of such books as Sir Richard Owen, "On the Nature of Limbs" [London, 1849], or Humphry, "On the Human Foot and Human Hand" [Cambridge and London, 1861]; but it will not, I think, be out of place if I note here some of the physiological features of this perfect member, which, as Aristotle says, is not part of another member, but is a perfect whole divided into parts like any other member.[22]

Owen.
Humphry.
Aristotle.

¶ 30.
The hand is a distinct member.

It will be my object to show that the hand is not a mere appendage, but is intimately connected with the entire frame; and to prove the statement of Sir Charles Bell, "That the hand is not a thing put on to the body like an additional movement to a watch, but that a thousand intricate relations must be established throughout the whole frame in connection with it; not only must appropriate nerves of motion and of sensation, and a part of the brain having correspondence with those nerves be supplied, but unless, with all this superadded organisation, a propensity to put it into operation were created, the hand would lie inactive."

¶ 31.
Composition of the hand.

The exquisite composition and mechanism of the hand has been summed up thus by Professor Sir Richard Owen.[23]: "The high characteristics of the human hand and arm are manifested by the subordination of each part to a harmonious combination of

---

[22] ARISTOTLE, ΠΕΡΙ ΤΑ ΖΩΑ ʹΙΣΤΟΡΙΟΝ, Βιβλ. ά., Κεφ. ά:—
"Ταῦτα γὰρ αὐτά τ' ἐστὶ μέρη ὅλα, καὶ ἔστιν αὐτῶν ἕτερα μόρια. Πάντα δὲ τὰ ἀνομοιομερῆ σύγκειται ἐκ τῶν ὁμοιομερῶν, οἷον χεὶρ ἐκ σαρκὸς καὶ νεύρων καὶ ὀστῶν.

[23] RICHARD OWEN, "On the Nature of Limbs. A discourse delivered on Friday, Feb. 9 [1849], at an evening meeting of the Royal Institution of Great Britain," p. 36 (London, 1849).

AN INTRODUCTORY ARGUMENT. 37

function with another, by the departure of every element of the appendage from the form of the simple ray, and each by a special modification of its own; so that every single bone is distinguishable from another; each digit has its own peculiar character and name; and the 'thumb,' which is the least important and constant of the five divisions of the appendage in the rest of the class, becomes in man the most important element of the terminal segment, and that which makes it a 'hand,' properly so called." [24]

¶ 32. *The bones of the hand.*

Let us take a rapid survey of the bones of the hand. At the ends of the radius and ulna [the two bones of the fore-arm] we find in the human skeleton the carpal bones, which, being eight in number [scaphoid, semilunar, cuneiform, pisiform, trapezium, trapezoid, magnum, and unciform], and fitted closely into one another, compose the *wrist*. Beyond these we have the five metacarpal bones, which supply the framework of the palm; and above these again, the three rows of phalanges which constitute the fingers and the two phalanges which constitute the thumb.

[24] Speaking of the feet, and contrasting the toes with the fingers, Professor Owen continues:—"In the pelvic, as in the scapular extremity, the same digit [the thumb], which is the first to be rejected in the mammalian series, becomes, as it were, the 'chief stone of the corner,' and is termed, *par excellence*, 'the great toe;' and this is more peculiarly the characteristic of the genus *Homo* than even its homotype the thumb; for the monkey has a kind of *pollex* on the hand, but no brute animal presents that development of the hallux on which the erect posture and gait of man mainly depend.' Galen, also, in his work "De Usu Partium corporis humani" [book ii.], remarks on this very point :—" Quare autem non, sicut digiti pedum, ex uno ordine siti sunt omnes, sed magnus est aliis oppositus, dictum quidem est et hoc, sed quantum deest nunc adjicietur. Pes quidem amubulandi organum fuit, manus autem apprehendi. Conveniebat autem illi quidem firmationis securitas; huic autem apprehensionis multiformitas. Sed formationis quidem securitas in uno ordine locatis indigebat omnibus digitis, promptitudo vero ad varietatem acceptionum, magno, id est pollice indiget aliis opposito."

Owen on the foot.

¶ 33.
Comparative anatomy.

The importance of a thorough comprehension of these bones and *their relationship to one another* is very clearly and most interestingly set forth by Professor Sir Richard Owen [Op. cit., p. 29], when, after pointing out the various relationships as exemplified by comparative anatomy, he says:—
"Another important and instructive result of the foregoing comparisons is the constancy of the relations of the distal series of carpal and tarsal bones, whether single or compound, with the five digits with which they essentially correspond in number; for by this constancy of connection we are able to determine the precise digits that are lost and retained when their number falls below the typical five; to point out, for example, the finger in the hand of man that answers to the forefoot of the horse, and the toe that corresponds with the hind foot; nay, the very nail which becomes by excess of development the great hoof of the horse." To illustrate this, we may quote the following passage:—"The small styliform ossicle which is attached to the *trapezium* in the wrist of the spider monkey (*Ateles*), or the hyæna, is plainly shown *by that connection* [besides its relations to the other digits] to be a remnant of the thumb . . . . . The similar ossicle that is attached to the diminished unciforme of the marsupial bandicoot, is shown *by that connection* to be the rudiment of the little finger, the three remaining digits also retaining respectively their normal connections with the trapezoïdes, the magnum, and the unciforme," etc., etc., etc.

¶ 34.
Muscles of the hand.

A full consideration of the muscles of the hand would take too much time for us to enter upon it here; shortly, as described by Humphry,[25] they are as follows:—"The wrist and hand are bent forwards or flexed upon the forearm by three mus-

[25] G. M. HUMPHRY, "The Human Foot and Human Hand" (Cambridge and London, 1861).

AN INTRODUCTORY ARGUMENT. 39

cles, which pass downwards from the inner condyle, [or expanded end of the humerus], and are termed the *radial flexor*, the *ulnar flexor*, and the *long palmar* muscles. The first two of these muscles are inserted into wrist bones on the radial and ulnar sides respectively, while the third expands into a fan-like *fascia*, or membrane, in the palm of the hand, and thus serves both to support the skin of the palm, and to protect the nerves and vessels which lie below it. Beneath the *palmar fascia* lie two sets of *flexor* muscles of the fingers, and they present so beautiful a mechanical arrangement as to merit special notice. The *superficial* or *perforated flexor* muscle passes down the front of the forearm, and divides into four tendons, which become apparent after the removal of the *palmar fascia*, and are inserted into the second phalanges of the fingers, each tendon splitting at its termination to give passage to the similar tendons of the *deep*, or *perforating flexor* muscle, which passes from the upper part of the ulna to be inserted in the last phalanx of each finger. These *flexor* muscles are antagonised by the *common extensor* muscle of the fingers, which, like the flexors, divides into four tendons, one for each finger. Besides these there is a special *extensor* of the index finger, a series of muscles forming the ball of the thumb, which move that digit in almost every direction, and various small slips, giving lateral and other movements to the fingers." The thumb is, in point of fact, better supplied with muscles than any of the fingers, and, as we shall see, it is this bundle of muscles which constitutes that most important part of the hand known in Cheiromancy as the Mount of Venus [*vide* ¶ 371], by the greater or less development of which we discover the greater or less strength of animal will of the subject under examination ; and it is an interesting and significant fact that

The thumb muscle.

by a contraction of the flexor, and non-development of the extensor muscles the human infant hides its thumb in the palm of its hands until its *will* shall have developed itself and put itself into exercise, and a test of complex "will organization" is to be found in an early and complete development of extensor action.

¶ 35.
Importance of the thumb.

We must not lose sight of the enormous importance of the thumb in the œconomy of the hand [*vide* Note [24], p. 37], an importance which is excellently expressed by Galen [Op. cit., book I.], when he says:—"For suppose, the thumb were not placed, as it is, in opposition to the other four fingers, but that all the five were ranged in the same line; is it not evident that in this case their number would be useless? For in order to have a firm hold of anything it is necessary either to grasp it all round, or at any rate to grasp it at two opposite points, neither of which would have been possible if all the five fingers had been placed in the same plane; but the end is now fully attainable, simply in consequence of the position of the thumb, which is so placed, and has exactly such a degree of motion as by a slight inclination to be easily made to co-operate with any one of the four fingers."[26] I have already quoted another passage in which Galen refers to this importance of the thumb. [Note [24], p. 37.]

Galen.

¶ 36.
Influence of the habits on the tissues, and of the tissues on the bones.

I have dwelt upon these anatomical details, because the form of the bones, their consistency, and the development of the muscles *depend almost entirely*

[26] "Quid namcque si nullus digitis quatuor, ut nunc habet, opponeretur, sed consequentur omnes quincque sub unâ rectâ lineâ essent facti? Nonne perspicuum est eorum tunc multitudinem fore inutilem? Quandoquidem quod tuto fideliterque apprehenditur, aut undique circulo, aut omnino ex locis duobus contrariis, comprehendatur oportet. Id quod periisset, si omnes sub unâ rectâ lineâ uno ordine facti fuissent digiti. Verumtamen hoc ipsum, digito uno aliis opposito, diligenter servatum est. Qui quidem positione et motu ita habet, ut parva omnino flexione curvatus, cum singulis quatuor oppositis actionem perficiat."

upon the habits and characteristics of the owner of the hand; and this is an important factor in my argument as to the physiological science of Cheirosophy, viz.: that, as certain habits and characteristics produce certain developments of bone and muscle, so from the appearances of those developments in a hand may the habits and characteristics of a subject be unmistakably inferred. That this is the case will not, I apprehend, be for one moment doubted; the texture of bone is essentially elastic, and, in examining a human skeleton, it is a recognized fact that the irregularities and prominent ridges found upon the surfaces of the bones are the results of the actions and pressures of the superincumbent mass of muscles, nerves, and venal plexus. "This explanation of the use of the prominent ridges of a bone," says Sir Charles Bell, "imparts a new interest to osteology. The anatomist ought, from the form of the ridges, to deduce the motions of the limb and the forces bearing upon the bone. It is, perhaps, not far removed from our proper object to remark that a person of feeble texture and indolent habits, has the bone smooth, thin, and light; but that nature, solicitous for our safety in a manner that we could not anticipate, combines with the powerful muscular frame a dense and perfect texture of bone, where every spine and tubercle is completely developed. And thus the inert and mechanical provisions of the bone *always bear relation to the living muscular power of the limb;* and exercise is as necessary to the perfect constitution and form of a bone as it is to the increase of the muscular power."

Sir C. Bell

As to the veins and arteries of the hand I should not say anything in this place, were it not that a very curious and ancient superstition, or rather custom, existed concerning the third or ring finger; a superstition founded in times of the remotest antiquity, upon an ignorance of anatomy which is both interest-

¶ 37.
Superstition as to the ring finger.

ing and curious, and which resulted [as we shall see] in this finger being termed *Medicus* [vide Note [112], p. 186]. In very old works on medicine we find directions to the effect that, in mixing their drugs, doctors are to use nothing but the third finger, this being connected directly with the heart by a main artery [?!], and the reason of this is thus stated by Levinus Lemnius:[27]

Lemnius.

"So I observed in *Gallia Belgica* that very many were subject to the gowt [*gout*] of their hands and feet, all whose joynts were swolln and in bitter pains, save onely the ring finger of the left hand which is next the little finger, for that by the nearnesse and consent of the heart felt no harm . . . . . Because a small branch of the arterie [and *not* of the nerves as Gellius thought] is stretched forth from the heart unto this finger . . . . . Also the worth of this finger that it receives from the heart procured thus much that the old Physitians [from whence also it hath the name of *Medicus*] would mingle their medicaments and potions with this finger, for no venom can stick upon the very outmost part of it, but it will offend a man and communicate itself to his heart;" he then goes on to explain that this leads to the circumstance of its being the *ring*-finger, and on this very matter of the ring-finger Kirchmann has some most interesting explanatory notes of which the following is a translation :[28]—After pointing out the fact of this being, *par excellence*, the ring-finger, and giving as authorities for some interesting remarks thereon, Pliny, Macrobius, and Politianus, he points out that the thumb has been discarded as a ring-finger, because of its comparative inferiority of formation, the second and fourth on account respectively of their largeness and smallness,

Kirchmann.

---

[27] LEVINUS LEMNIUS, "The Secret Miracles of Nature in four Books" (London, 1658), bk. ii., ch. 11.

[28] JOHANNIS KIRCHMANNI, Lubecensis, "De Annulis liber singularis" (Slesvigæ, 1657).

and that, closed in as it is on either side, the third becomes eminently the suitable finger. The Ægyptians [says he, quoting Disarius and Macrobius], being eminent anatomists, discovered the presence of a nerve from the heart to this finger, and he points out that Agellius supports the statement. Then follows a quotation, "Johannes Salisberiensis libro vi; Policratici cap. xii; Gratianus, Can. *Fœmine* xxx, quæst. v; ex Isidori Hispalensis lib. ii. De Divinis Officiis, cap xix," to the effect, that a *vein* makes this connection; and then he says, " But this opinion is now exploded by modern doctors, who, after careful dissection of the human body, *have found no artery or vein* extending thus from the heart to this finger, and there is no such reason for the preference of this finger for wearing the ring."

For a concise and accurate description of the arterial and venous system of the hand I refer the reader to F. T. McDougall's article on the subject in R. B. Todd's "Cyclopædia of Anatomy and Physiology," vol. ii., 1836-9. The hand contains two principal arteries, the *radial* and the *ulnar*. The *ulnar* proceeds in a curve from the wrist to the first finger [where it joins a branch of the *radial*], forming what is known as the *palmar arch*. Four digital arteries go from its convexity, which subdivide into collateral branches about two lines below the metacarpo-phalangean articulations [*vide* ¶ 32]. These supply the palmar and lateral surfaces of the fingers, excepting the thumb and the outer side of the index, the branches coalescing at the tips of the fingers, whence branches arise to supply the pulp of the fingers with blood. The *radial* goes from the end of the forearm [radius] round the wrist, into the thumb, and back into the palm, where, joining the *ulnar* [as I have said above], it forms what is known as the *deep palmar arch*. Before, however, it does this, it gives off two veins,

¶ 38. Arteries. Venous system of the hand.

Ulnar artery.

Radial artery

one of which [the *superficialis volæ*] supplies the palm, whilst the other divides into two branches [the *arteriæ dorsales pollicis*] which run along either side of the thumb, one of which branches sends off an artery to the index-finger. When the *radial* dips into the palm it gives off branches to the thumb and fore-finger, or index, and to the deep palm [where it joins the *ulnar*]. The arrangement of these arteries varies in different hands, a significant fact for us, which is noticed in another place [*vide* ¶ 72]; the veins, which are generally very deep, accompany the arteries; there are very few superficial veins in the hands. These light notes will, I think, assist our anatomical comprehension of the hand, and will help to impress upon us the obvious *design* which enters into its construction, so that we may say with Professor Owen :—" With regard to the structural correspondences manifested in the locomotive members, if the principle of *special adaptation* fails to explain them, and we reject the idea that these correspondences are manifestations of some architypal exemplar on which it has pleased the Creator to frame certain of His living creatures, there remains only the alternative that the organic atoms have concurred *fortuitously* to produce such harmony,"—an Epicurean argument [29] from which every healthy mind naturally recoils.

¶ 39. Nervous system of the hand. Of course, the most important subject for our consideration in this place is that of the nervous system of the hand, of that complicated plexus of nerves which gives to the hand its direct and constantly apparent connection with the brain. There are more nerves in the hand than at any other point of the body,[30] and in the palm they are more numerous than at any other point of the hand. It is this that

---

[29] 'Ἀπὸ τῶν ἀτόμων σωμάτων, ἀπρονόητον καὶ τυχαίαν ἐχόντων τὴν κίνησιν.—EPICURUS, " Physica et Meteorologica " (J. G. Schneider : Leipsic, 1813).

[30] Aristotle calls attention to this in his ΠΕΡΙ ΤΑ ΖΩΑ ΊΣΤΟ.

causes the feeling of revulsion and of sickness which ensues, when the palms of the hands or the soles of the feet are tickled; it is by reason of this that in fever the hands become burning hot, whilst the rest of the body, which is more muscular than nervous [as opposed to the hands, which are more nervous than muscular], is cold; and the hands and feet become numbed by cold or fear sooner than the rest of the body, by reason of the high development of the vaso-motor nerve arrangements in them, added to the circumstances of their rich blood-supply, distance from the heart and delicate skin covering, thus showing that the hand acts as the thermometer, so to speak, of the life. *Elementary indications.*

Without the hand, principal seat as it is of the SENSE OF TOUCH, the other senses would be comparatively useless; the sense of Touch is the only sense which is reciprocal. [In the sense referred to by Sir Walter Scott, where he alludes to the sensations produced by touching one's own body unconsciously.[0a]] That is, though the senses of Sight, Hearing, Taste, and Smell, can only *receive* impressions without giving them, that of Touch both *receives* and *gives;* and it is this sense of Touch, dependent as it is upon the nervous system, which is the most important of all, and which is found in its highest state of development where that nervous system is the most complete, namely, in the hand. *¶ 40. The sense of touch. Its superiority.*

I cannot, I think, do better than follow the principles upon which Bernstein discusses the physiology of the sense of Touch, as an introduction to this section of my argument.[81] Every sensory organ may be shown *¶ 41. Physiology of the sense of touch. Bernstein.*

PION, where he says, Βιβλ. Γ'., Κεφ. ἑ. :—"Πλεῖστα δ'ἐστι νεῦρα περὶ τοὺς πόδας καὶ τὰς χεῖρας καὶ πλευρὰς καὶ ὠμοπλάτας καὶ περὶ τὸν αὐχένα καὶ περὶ τοὺς βραχίονας."

[0a] SIR WALTER SCOTT, "Letters to J. G. Lockhart on Demonology and Witchcraft" (London, 1830), Letter 1.

[81] JULIUS BERNSTEIN, "The Five Senses of Man" (London, 1883), 4th Edition.

to be anatomically connected with the nervous system by means of nerve trunks and nerve fibres. Touch, sight, hearing, smell, and taste are inconceivable without the presence of a nervous system, even if the sensory organs were in their present full development [*e.g.*, an arm of which the nerve is injured can feel nothing], the *sensation* itself evidently first takes place in the brain [*e.g.*, the sensation of *light* does not take place in the eye, where there is only an impression of light upon the expanded surface of the optic nerve; the sensation of light cannot, however, take place in the optic nerve itself, for it merely conveys the fact of the existence of the irritation to the brain]. Of all the senses the most perfect is undoubtedly that of Touch, and though it is very greatly assisted by that of Sight, still, the former can dispense with the latter far better than the latter with the former. The simultaneous action of the sensations of touch and sight is, in fact, for the human mind an important source of knowledge in the external world. Yet we must not on this account conclude that the sense of Touch alone, without the assistance of sight [as in the case of persons born blind], cannot lead to knowledge. It is probable that the sense of touch alone might enable us to distinguish our own body and external objects *sooner* than vision. For the act of touching our body with our hand calls forth a *double* sensation of touch, one through the hand, and the other through the part of the skin touched, whilst touching an *external object* causes only a single sensation of touch through the tactile organ.

¶ 42.
The sensation of pain.

And it must be remembered that the sense of Touch is our great bodily safeguard, for it produces the sensation of pain [as distinguished from that of contact], which warns us to fly from the agent which produces that sensation. The limit between the sensations of touch and pain may be illustrated by the following example given by Ernst Heinrich Weber.[32] If we place the edge of a sharp knife on the skin, we feel the edge by means of our sense of touch; we perceive a sensation, and refer it to the object which caused it. But as soon as we cut the skin with the knife we feel pain, a feeling which we no longer refer to the cutting knife, but which we feel within ourselves, and which communicates to us the fact of a change of condition in our own body; by the sensation of pain we are not able to recognise with the same degree of accuracy either the object which caused it, or its nature. *Touch and pain contrasted.*

Let us then examine this sense of Touch, and particularly let us examine it with regard to the part which the hand plays in its development. We know from experience that every part of our skin possesses a certain sensibility, and that this sensibility varies in different parts. This property is given to the skin by a great number of nerves which originate in the brain and spinal cord, and extend in a tree-like form over the body. The sensibility of any part of the body is due to these nerves alone, for, as soon as such a nerve is lost or diseased, the part of the body supplied by it becomes void of sensation. ¶ 43. *Anatomy of the sense of touch.*

And this sense of touch may be said to be the only universal sense; for, as Aristotle and Cuvier have both remarked in parallel passages which are quoted by Dr. Kidd [*vide* note [4], p. 21], that of the five senses, touch alone is common to all animals, and is so generally diffused over the whole body that it does not [like the other senses] reside in any specific part *alone*. All animals do not possess *all* the senses some possess only a part of them; but no animal is ¶ 44. *The universal senses Aristotle. Cuvier.*

[32] ERNST HEINRICH WEBER, "De pulsu, resorptione auditâ et tactu annotationes anatomicæ et physiologicæ" (Lipsiæ, 1834), 4to.

without the sense of touch.³³ And the sense of touch does not only determine size, shape, and pressure; it alone of *all the senses* can appreciate differences of temperature, heat and cold.

¶ 45.
Touch and temperature.

The two great senses, therefore, which reside in the skin, are those of *touch* and of *temperature*. In touching a body we employ the organs best adapted for the purpose, namely the hands; and we can recognise the object touched with closed eyes, with more or less certainty; in the hands this power is very perfect, and is the more perfect the nearer we approach to the tips of the fingers, where the skin is the most sensitive.⁸⁴

¶ 46.
The skin.
Its composition.

The skin itself consists of three layers. Upon the cellular tissue *under* the skin lies the *first skin* [or *dermis*], which is of a tolerably compact nature. Its surface consists of a greater or less number of cylindrical or conical protuberances, which are called *papillæ*. Upon the dermis lies the *mucous layer*, which consists of a great number of small microscopic cells,

---

³³ ARISTOTLE, ΠΕΡΙ ΤΑ ΖΩΑ ἹΣΤΟΡΙΟΝ, Βιβλ. Δ΄., Κεφ. ή.:—
" Εἰσὶ δ᾽ αἱ (αἰσθήσεις) πλεῖσται καὶ παρ᾽ ἃς οὐδεμία φαίνεται ἴδιος ἑτέρα, πέντε τὸν ἀριθμὸν, ὄψις, ἀκοὴ, ὄσφρησις, γεῦσις, ἁφή. ... Οὐ γὰρ ὁμοίως πᾶσιν ὑπάρχουσιν, (αἰσθήσεις) ἀλλὰ τοῖς μὲν πᾶσαι τοῖς δ᾽ἐλάττους. Τὴν δὲ πέμπτην αἴσθησιν τὴν ἁφὴν καλουμένην καὶ τ᾽ἄλλα πάντ᾽ ἔχει ζῷα." " Πᾶσι δὲ τοῖς ζῴοις αἴσθησις μία ὑπάρχει κοινὴ μόνη ἡ ἁφὴ ὥστε καὶ ἐν ᾧ (αὕτη) μορίῳ γίνεσθαι πέφυκεν ἀνώνυμόν ἐστίν." Βιβλ. Α΄., Κεφ. γ΄. And the parallel passage of Cuvier runs:—" Lesens extérieur le plus général est le toucher, son siège est à la peau, membrane enveloppant le corps entier. ... Beaucoup d'animaux manquent d'oreilles et de narines; plusieurs d'yeux : il y en a qui sont reduits au toucher *lequel ne manque jamais.*"—G. CUVIER, "Le Règne Animal distribué d' après son organization" (Paris, 1828).

³⁴ To those of my readers who desire to go deeper into the relationship between the sense of Touch and the hand, I would recommend Dr. Arthur Kollmann's work "Der Tast-apparat der Hand der menschlichen Rassen und der Affen in seiner Entwickelung und Gliederung" (Hamburg und Leipzig, *L. Voss;* 1883)

AN INTRODUCTORY ARGUMENT. 49

completely filling the depressions between the papillæ of the dermis. Lastly, the outer layer, the *cuticle* [or *epidermis*], which consists of an intergrowth of cells which are filled with a solid horny substance. The blood-vessels and nerves only extend as far as the surface of the dermis and to its papillæ; the mucous layer and the epidermis are completely free from blood and nerves. The nerves of the skin, which terminate in single fibres, only extend to the dermis, and here they are observed to end in a peculiar manner in the papillæ. Many of them contain, for instance, an egg-shaped particle, which a nerve fibre enters, and in which it is lost after several convolutions round it. They are called *tactile corpuscles*, and there can be no doubt that they act as the instruments of the sensation of touch. They are not found in the same numbers in all parts of the skin, occurring in the greatest number in those parts where the sensibility is more acute, and more sparingly where weaker. *[Tactile corpuscles.]*

They are extraordinarily numerous at the tips of the fingers, where, in the space of a square line, about a hundred can be counted, they are tolerably numerous over the whole surface of the hand, but occur in much smaller numbers on the backs of the hands.[35] On the palm of the hand also, the papillæ [which, however, do not *all* contain a tactile corpuscle] occur in great numbers, *and are arranged in regular rows; it is these rows of corpuscles which cause the lines in the hands* of which we shall have so much to say further on: let this, therefore, be borne in mind. The nerves of the skin and deeper parts are observed also to possess another terminal apparatus, similar to that of *[¶ 47. Distribution of the tactile corpuscles.]* *[The lines in the palm.]*

---

[35] "Mira vallecularum tangentium in interna parte manus pedisque, præsertim in digitorum extremis phalangibus, dispositio flexuræque attentionem jam nostram in se trahit."—JAN. E. PURKINYE, "Commentatio de Examine Physiologico," etc. (Lipsiæ, 1830).

Pacinian bodies. the tactile corpuscles, namely, long globules [pacinian bodies], in the hollows of which the nerve fibres terminate. In short, in the entire surface of the skin there exist terminal apparatus of a peculiar kind for the sensory nerves, and if we wish to follow the action of sensation further physiologically, we must start with the excitement of a nerve fibre which ends *Connection with* in a definite part of the skin, and follow the course of *the brain.* the excitement to the brain. The course of the nerve between brain and skin, along which the excitement passes, can be followed anatomically with a certain degree of exactness. A nervous fibre which ends in the skin forms, as far as its union with the spinal cord or brain, a long fine continuous thread. The fibres which terminate in the skin very soon unite in small branches, and finally in thick nerve trunks, before they enter the central organ of the nervous system, *but* [practically] *in no case do two nervous fibres coalesce in these nerve branches.* We may, therefore, assume that *every* part of the skin is provided with *isolated connections* with the centre of the nervous system, which are united there just as telegraph lines unite at a terminus.

¶ 48 Julius Bernstein is not the only writer who has
Parallels be- likened the nervous system to an electric telegraph,
tween the
nervous system and undoubtedly it is an extremely happy simile, for,
and electricity. the brain being the fountain head of the life, without it we should be dead to all impressions, physical or mental; thus, if a blow strikes the leg, the nerves there terminating, instantly carry to the brain the intelligence of a blow having been received, and this communication constitutes the sense of feeling; if the nerve is divided between the leg and the brain there is no telegraph to convey the news, and the brain, the sensitive power, has no intelligence of the blow, which consequently does not hurt the recipient.[36]

[*] M. Desbarrolles, in the fifteenth edition of his work already cited [note [a], p. 67], gives a striking and absolute proof of this

A1. INTRODUCTORY ARGUMENT. 51

Abercrombie remarks in this connection,[37] "The communication of perceptions from the senses to the mind has been accounted for by motions of the nervous fluid, by vibrations of the nerves, or by a subtle essence resembling electricity or galvanism;" and Müller, in his "Physiologie" says, "Perhaps there exists between the phenomena of the nervous system and of electricity a sympathy, or connection at present unknown, analogous to that which has been found to exist between electricity and magnetism. The one thing which the recognised march of enlightenment forbids us is the employment of a conjecture which reposes upon nothing in the construction of a scientific system;" and in another place he says:—"We know not *as yet* whether or no, when the nerves convey an impression, an imponderable *fluid* flies along them with an inconceivable rapidity, or whether the action of the nervous system consists of the oscillations of an imponderable *principle*

¶ 49.
Abercrombie

Müller.

fact, p. xxix). A gentleman, having called upon him for a cheiromantic consultation, and having submitted to M. Desbarrolles his left hand, he was told of an accident which had happened to his other arm and its concomitant circumstances. The gentleman then uncovered his right arm which up to that time had been concealed, saying as he did so that the nerves having been destroyed it was absolutely dead to him, and that he could neither move it nor feel anything which touched it. M. Desbarrolles looked at the palm thus deprived of its sensitive power, and found, as might have been anticipated, that the lines which depend [as Bernstein says very truly; *vide* ¶ 47] upon the tactile corpuscles of the nerves had absolutely disappeared. If it were not that these lines depended for their existence on the nerves, and the imponderable nerve fluid to which we shall next animadvert, the hand would indeed have been dead and insensible, but the lines already formed would have remained; as it was, they had disappeared with the nerves, with the fluid by whose agency they had been formed, and on whose presence they depended for their existence.

[37] J. ABERCROMBIE, "Inquiries concerning the Intellectual Powers and the Investigation of Truth," 9th ed. (London, 1838)

already existent in the nerves and placed in vibration by the brain": I will pray the reader to bear this theory in mind, as I shall have occasion to refer to it again. The philosopher Herder,[38] also paid great attention to the action in our lives of this imponderable, sensitive fluid, which he likens to electricity, but for which he claims a vast superiority over electricity as a more subtle, more sensitive essence, to which we owe our lives, and which for want of a better term is known to us as our *soul*.

¶ 50. Sir C. Bell's classification of nerves.

We must also bear in mind the discovery first made by Sir Charles Bell, that nerves are divided into two classes, *sensory* and *motor*, and he carries out what I have said above in the following passage, which occurs in his Bridgewater Treatise "On the Human Hand":—"It was the conviction that we are sensible of the action of the muscles which led me to investigate their nerves; first, by anatomy, and then by experiment. I was finally enabled to show that the muscles are provided with two classes of nerves; that on exciting one of these the muscle contracts; on exciting the other no action takes place; *and that the nerve which has no power over the muscle is for giving sensation.* Thus it was proved that muscles are connected with the brain through a 'nervous circle'; that one nerve is not capable of transmitting what may be called *nervous influence* in two different directions at once; in other words, that a nerve cannot carry volition to the muscles and sensation to the brain, simultaneously, and by itself; but that for the regulation of muscular action two distinct nerves are required; first, a nerve of sensibility to convey a consciousness of the condition of the muscles to the sensorium; and secondly, a nerve of motion for conveying a mandate of the will to

Sensory and motor nerves.

---

[38] HERDER, "Idées sur la Philosophie de l'Histoire de l'Humanité" (Paris, 1827).

AN INTRODUCTORY ARGUMENT. 53

the muscles." In this manner Sir Charles Bell showed that every apparently simple nerve is in reality compound, consisting of two nerves springing from different roots, but enclosed in a single sheath, one conveying volition from the brain to the tissues, and one conveying impressions from the tissues to the brain ; and it is this double function of the duplicate nerves which gives to the sense of touch the superiority claimed for it above of being active and passive [vide ¶¶ 40 and 41].

We must also observe before leaving this branch of the subject that the sense of touch is keenest where the vascularity is the highest [i.e., where the blood-vessels are most numerous, the nervous plexus is nearest to the surface], that is, of course, where the skin is the most red ; as, for instance, at the lips, at the rosy tips of the fingers, and along the pink lines of the hand. "From the constant call for vigorous and rapid, as well as sustained and powerful action," says McDougall,[39] "the hand, with the exception of the tongue, is the most vascular of the voluntary locomotive members of the human body." Hunter observed that a distribution of nerves and tactile corpuscles to any part invariably carries with it an increased vascularity, so that, where the nerves are most numerous, the blood-vessels are also most finely distributed, and there the sense of touch is keenest ; and Sir Charles Bell observed, in confirmation of this, that the human infant first developes the sense of touch in the lips and tongue, the *next* motion being to put its hand to its mouth, and as soon as the fingers are capable of grasping anything, whatever they hold is carried to the mouth for further examination and identification. "The Hand," says he, "destined to become the instrument for perfecting the other senses, and for developing the endowments of the mind itself, is in

¶ 51.
Effect of vascularity on the sense of touch.

Dr. Hunter.

Sir C. Bell.
The human infant.

* In R. B. Todd's " Cyclopædia of Anatomy and Physiology " (London, 1836-9), vol. ii.

the infant absolutely powerless. There occur certain congenital imperfections in early childhood which require surgical assistance, but the infant will make no direct effort with its hand to repel the instrument, or disturb the dressing, as it will do at a somewhat later period." This is a very interesting illustration of the gradual development of the sense of touch, commencing at points of the highest vascularity.

¶ 52. Indications of disease. Finally, let us remember the indications of disease which are afforded by the hands, indications which [unless we are in error] emanate directly from the brain by means of this nerve-communication of which we have said so much. "Cutaneous phenomena," says **Georget.** Georget,[40] "*regulated by the influence of the brain,* though they are less clearly evident than some others, are none the less real and worthy to fix the attention of the observer." I have before called attention to the **Temperature.** manner in which the temperature of the hands announces indisposition: the signs of various illnesses which we find in the hands, and which are all noticed in their proper places during the course of this work, are numerous and convincing.

¶ 53. Further Indication Not to multiply instances, we may cite the symptom of approaching leprosy, which is found in the stiffening of the first finger, and the fact that among scrofulous persons we always find a thick first finger and short nails. Filbert nails, as a sign of a tendency to consumption, are well known, and always recognized.

¶ 54. We must, I am afraid, dismiss, with other similar superstitions of the old doctors, the statement, made **Avicenna.** by Avicenna, that short fingers indicate weakness of the liver.[41]

---

[a] GEORGET, "Physiologie du Système Nerveux" (Paris, 1821): —"Les Phénomènes cutanés *déterminés par l'influence cérébrale,* quoique moins évidents que dans plusieurs organes, sont pourtant réels et dignes de fixer l'attention de l'observateur."

[41] AVICENNÆ, "Liber Canonis de Medicinis Cordialibus,"

AN INTRODUCTORY ARGUMENT.

Now, therefore, that we have arrived at a sufficiently thorough comprehension of the member with which we are concerned, let us turn to the consideration of the history and progress of the science of Cheirosophy, and of its aims and objects as an inducement to its study.

Cheiromancy, as a science, is one which we find continually referred to by classical and other authors, and many modern and ancient writers on the science have claimed for it the sanction and authority of Scripture, quoting various passages of Holy Writ in support of their statements, but notably one, concerning which so much has been written that I think it not unwise to dissect its claims to such an interpretation in this place at some length and with some completeness. This most important and universally quoted text occurs in the thirty-seventh chapter of the Book of Job, verse 7, and runs in our English version: "He sealeth up the hand of every man; that all men may know His work." ¶ 55.
Cheiromancy and the Holy Scriptures.

Now, in the Vulgate this passage reads, "*Qui in manu omnium hominum signat ut noverint singuli opera sua*"; and author after author has reproduced this passage as referring to Cheiromancy, a treatment which has, perhaps not unnaturally, given rise to some considerable amount of discussion. The commentator Valesius took this passage to mean that the signs which are placed upon man's hands, were placed there for his instruction, and this reading has, of course, been freely adopted by the Cheiromants; but by far the greater number of authors take the passage to signify that the snow and rain mentioned in the pre- ¶ 56.
The Vulgate

Valesius.

translated into Latin by Gerardus Cremonensis (Venice, 1582). Liber III., tract. i., cap. 30 (*De signis*), cap. 29 (*De parvitate hepatis*) :—"Et assiduatur debilitas digestionis et multiplicatur eventus oppilationis, et apostematum *et certificat illud brevitas digitorum in creatione*."

ceding verse being poured upon the earth, the hand of man is, as it were, "sealed up," and during his enforced inactivity his thoughts turn naturally to higher things. Mercerus [42] takes the view that this shutting up, as it were, of the hands by rain, is a sign to men that they are but created beings, subject to a will higher than their own. G. H. A. von Ewald and Albert Barnes [43] follow Mercerus, and the latter expressly mentions the pretensions of the Cheiromants, and points out their fallacy.[44] Calvin himself [45] remarks upon the reading, declaring that the text will not and cannot bear this interpretation; and Joseph Caryl in the next century,[46] after interpreting the passage in the agricultural sense of enforced inactivity, says:—
"The Hebrew is, 'In the hand He will seal,' or 'sealeth every man'; from which strict meaning some have made a very impious interpretation of this text, thereupon grounding that (as most use it) most unwarrantable art of chiromancy, as if God did put certain lines, prints, or seals, upon the hand of every man, from whence it may be collected and concluded what [as some call it] his fortune or destiny will be in the world, which, as it is an opinion wicked in itself, so altogether is heterogeneal to this place," etc. Hutcheson [47] gives the interpretation

[42] JOANNIS MERCERI . . . . "Commentarii in Jobum" (Amsterdam, 1564), p. 298.
[43] ALBERT BARNES, "Notes Critical, etc., on the Book of Job" (Glasgow, 1847).
[44] "Alii sensus ab hoc loco sunt alieni, ut quod chiromantici ex priore hemistichio suam artem colligunt, signasse Deum in manu hominum et notas ac signa impressisse unde res cognoscantur, etc. Quod eo nihil facit et extra rem est."
[45] "Sermons de M. Jean Calvin sur le livre de Job" (Genève, 1563).
[46] JOSEPH CARYL, "Exposition on Job" (London, 1664), vol. xi.
[47] G. HUTCHESON, "Exposition on the Book of Job" (London, 1669).

AN INTRODUCTORY ARGUMENT. 57

above referred to, *i.e.*, that " .... by these storms He *sealeth up the hand of every man*, or hindereth their work abroad, as if their hand were shut up under a seal. ... It should be men's special exercise, when they are restrained from their callings, to study to know God and His works well," etc. On the other hand, Schultens [48] seems to favour the reading of the passage in a manner favourable to the art of cheiromancy [for it could not in those days be called a science], quoting the like opinions of the theologians Lyrannus, Thomassin, Delrio, and Valesius; but even he inclined to the opinion of the Jesuit Pineda, which is given below. Earlier in the same century, Sebastian Schmidt [49] cites the agricultural reading as the only possibly correct one, treating the cheiromantic interpretation with contempt.[50] Pineda has devoted a vast amount of work to the unravelling of this point, and, after recapitulating nearly everything which had been written thereon up to his day, gives a long section headed " Physiognomica et Chiromantica, interpretationes duæ," in which he gives some most interesting notes.[51] He cites the reading of Valesius, and then calls attention to many of the extracts I have given above from the works of Galen and Aristotle, fter which he balances all the *pros* and *cons* of the Cheiromantic interpretation, making the passage a peg on which to hang a somewhat lengthy dissertation on Cheiromancy. The commentator is of opinion

Schultens.

Schmidt.

De Pineda.

[48] ALBERTUS SCHULTENS, " Liber Jobi et Commentario Perpetuo," etc. (Leyden, 1737).

[49] SEBASTIAN SCHMIDT, " In Librum Jjobi Commentarius," etc., 1705, p. 1390.

[50] " Nihil aliud significatur quam quod homines per nivem et pluvias vehementes sic se domi continere cogantur, si Deus velit, ut nihil operis in agris peragere queant .... Ineptiunt profecto qui hinc cheiromantiam exculpere conantur."

[51] JOANN. DE PINEDA, Societatis Jesu, " Commentariorum in Job tomus alter " (Venice, 1704).

that the character and propensities of men are more likely to be inscribed on their faces than on their hands, and as to this particular passage concludes that the agricultural meaning is most likely to be the right one. Finally [not to multiply authorities], in a comparatively recent edition of the Commentaries of Nicolaus de Lyra,[52] the editor lays great stress upon the agricultural reading in a foot-note, objecting very explicitly to the interpretation of Franciscus Valesius, notwithstanding that the author in the text supports the cheiromantic view, quoting many cited examples in support thereof.

¶ 57.
Divergence of opinions.

I will not weary the reader with further discussion of this time-honoured passage; opinions appear to be [very justly] divided, for though our English translation seems certainly to favour the agricultural reading, the original Hebrew, the Vulgate, and many other versions of the Book of Job, seem strongly to favour the cheiromantic; it is a point which each student will probably solve for himself. I have endeavoured in the above citations and notes to supply him with the necessary materials for arriving at that solution.[53]

[52] "Scripturæ Sacræ cursus completus," annotavit et edidit J. P. M. (Paris, 1841).

Other passages of Scripture cited in support of Cheiromancy.

[53] Other passages of Holy Writ have been quoted by various authors in works on Cheiromancy; none of them are, however, credited with these interpretations by Commentators, possibly on account of the distrust and ill-favour with which cheiromancy has always been regarded by the Church, disfavour of which I shall presently say more. For the interest of such readers as care to investigate these matters for themselves, the following are some of the principal passages, with references to commentaries which touch upon the question involved:—Proverbs iii. 16, "Length of days are in her right hand, riches and honour are in her left," *vide* Dr. Martini Geieri, "Commentaria in Proverbia" (Amsterdam, 1696); "Commentarii in Proverbia Salomonis," authore Thoma Cartwrighto (Leyden, 1617); "Commentarie upon the whole Booke of Proverbes of Solomon" (London,

AN INTRODUCTORY ARGUMENT.   59

Leaving, therefore, the question of whether or no the science of Cheirosophy is countenanced by Holy Writ, let us see what classic and modern writers have given to it their weighty consideration. We all know, of course, the much-quoted lines of Juvenal, which have been translated by Dryden:—

¶ 58.
Classic authors

Juvenal.

  "The middle sort, who have not much to spare,
  "To Chiromancer's cheaper art repair,
  "Who clap the pretty palm to make the lines more fair;"[54]

1596); Schultens [*vide* Note *, p. 57]; Commentarii in Jobum et Salomonis Proverbia Joannis Merceri" (Amsterdam), 1651.— 1 Samuel xxvi, 18, "What evil is in mine hand," *vide* Schmidt, "In Librum priorem Samuelis commentarius" (Argentorati, 1687).—Job xxxiv. 20, "The mighty shall be taken away without hand," *vide* Samuel Cox, "Commentary on the Book of Job" (London, 1880).— Habbakuk iii. 4, "God came from Teran . . . he had horns coming out of his hands and *there* was the hiding of his power," *vide* "Commentarius in Habacucum," auctore Theodoro Scheltinga (Leyden, 1748); William Green, "New Translation of the Book of Habbakuk" (Cambridge, 1755).—Exodus xiv. 8, "Israel went out with a high hand," *vide* "Commentary on the five books of Moses," by Richard, Lord Bishop of Bath and Wells (London, 1694); "Commentarius in Pentateuchum Mosis," auctore Cornelio a Lapide (Antwerp, 1714); "Annotations upon the five Books of Moses," by H. Ainsworth (London, 1627); "Joannis Calvini in V Libros Mosis commentarii," 1595; Andrew Willet, "Hexapla in Exodum" (London, 1633).—Colossians iv. 18, "The salutation by the hand of Paul," *vide* "A Commentarie of M. J. Calvine upon the Epistle to the Colossians," translated by R. V." (London, 1581).— Revelations xiv. 9, "And receive his mark in his forehead, and in his hand," *vide* M. Stuart, "A Commentary upon the Apocalypse" (Edinburgh, 1858); Daubuz, "Commentary on the Revelations" (London, 1720); "Epitome of Commentaries on the Revelations," by Hezekiah Holland (London, 1650).

[54] " Si mediocris erit, spatium lustrabit utrimque
    "Metarum, et sortes decet; frontemque manumque
    " Præbebit vati crebrum poppysma roganti."
—JUVENAL, " Sat.", vi., 581.
See also the alternative verse and literal translation given in NUTTALL AND STIRLING'S "D. Junii Juvenalis Satiræ" (London. 1836).

Aristotle. and we have constantly made reference to Aristotle's frequent citations of the Art during the course of this introductory argument. His work ΠΕΡΙ ΖΩΩΝ ΜΟΡΙΩΝ [Βιβλ. Δ'., Κεφ.ί] especially teems with such references. It is said that Aristotle, when travelling in Ægypt, found an Arabic treatise on this science of the hand, graven in letters of gold, upon an altar dedicated to Hermes, and that he sent it to Alexander, as being a study worthy the attention of the highest s*ç*avants, where it was translated into Latin by one Hispanus. An extremely early MS. of this work is preserved in the British Museum, and I have before me a black letter opusculum of twenty-two leaves [without pagination] entitled, "Chyromantia Aristotelis cum figuris" (Ulme, 1490]. In his "History of Animals" he gives a most interesting description of the various parts of the hand, and calls attention to the fact that those people who have two lines crossing the entire hand are long-lived, whereas those whose hands are not entirely crossed by these two lines are short-lived, referring doubtless, as we shall presently see, to the lines of Head and of Life; [55] and later he puts a Problem on this very point.[56]

¶ 59.
Production of the earliest printing press. In any case, whatever may be the antiquity of the science, it is interesting and encouraging to know that almost the first book that was printed, even before

---

[55] ARISTOTLE, ΠΕΡΙ ΤΑ ΖΩΑ ΊΣΤΟΡΙΟΝ, Βιβλ· Α'., Κεφ. ιέ.:
—"Χειρὸς δὲ θέναρ, δάκτυλοι πέντε. δακτύλου δὲ τὸ μὲν καμπτικὸν κόνδυλος, τὸ δ' ἄκαμπτον φάλαγξ. Δάκτυλος δ' ὁ μὲν μέγας μονοκόνδυλος, οἱ δ' ἄλλοι δικόνδυλοι. Ἡ δὲ κάμψις καὶ τῷ βραχίονι καὶ τῷ δακτύλῳ ἐντὸς πᾶσιν· κάμπτεται δὲ βραχίων κατὰ τὸ ὠλέκρανον. Χειρὸς δὲ τὸ μὲν ἐντὸς θέναρ, σαρκῶδες καὶ διῃρημένον ἄρθροις, τοῖς μὲν μακροβίοις ἑνὶ ἢ δυσὶ δι' ὅλου, τοῖς δὲ βραχυβίοις δυσὶ καὶ οὐ δι' ὅλου. Ἄρθρα δὲ χειρὸς καὶ βραχίονος καρπός. Τὸ δὲ ἔξω τῆς χειρὸς νευρῶδες καὶ ἀνώνυμον."

[56] ΠΡΟΒΛΗΜΑΤΩΝ ΛΔ', ί:—"Διὰ τί ὅσοι τὴν διὰ χειρὸς τομὴν ἔχουσι δι' ὅλης, μακροβιώτατοι; Ἢ διότι τὰ ἄναρθρα βραχύβια καὶ ἀσθενῆ."

AN INTRODUCTORY ARGUMENT. 61

movable types were used, was on the subject of Cheiromancy, that the inestimably valuable "block book," "Die Kunst Ciromantia," written by Hartlieb [whose portrait Miss Horsley has given us as our frontispiece], was written in 1448, and printed at Augsburg in the year 1475.[57] *Hartlieb*

Aristotle was not the only classical author who recognized the important functions of the hand. Quintilian refers to its expressive powers in several places, both as regards the use made of it by deaf and dumb persons, and as regards the multiplicity of things that may be expressed by its means, notably in the passage where he says :—" For though many parts of the body assist speech, *the hands* may be said actually to speak themselves, for do we not with the hands demand, entreat, call, dismiss, threaten, abhor, fear, interrogate, deny," etc.;[58] and it is probably from this passage that Montaigne in his " Apologie de Raimond Sebond " derives his celebrated passage :—" What about hands ? We request, promise, call, dismiss, threaten, entreat, supplicate, deny, refuse, interrogate, *¶ 60. Quintilian. Multiplied uses of the hand. Montaigne.*

[57] JOHANN HARTLIEB, " Die Kunst Ciromantia," printed at Augsburg, 1475.
[58] " M. FABII QUINCTILIANI " De Institutione Oratoria Libri duodecim " (Oxford, 1693), lib. XI., cap. iii. " De Gestu in Pronunciando," p. 578. " Quippe non manus solum sed nutus etiam declarant nostram voluntatem, et in mutis pro sermone sunt," p. 581. " Manus vero, sine quibus trunca esset actio ac debilis, vix dici potest quot motus habeant, cum pene ipsam verborum copiam persequantur : nam ceteræ partes loquentem adjuvant, hæ (prope est ut dicam) ipsæ loquuntur. An non his poscimus ? pollicemur ? vocamus ? dimittimus ? minamur ? supplicamus ? abominamur ? timemus ? interrogamus ? negamur ? gaudium tristitiam, dubitationem, confessionem, pœnitentiam, modum, copiam, numerum, tempus ostendimus ? Non eadem concitant ? supplicant ? inhibent ? probant ? admirantur ? verecundantur ? Non in demonstrandis locis atque personis, adverbiorum atque pronominum obtinent vicem ? ut in tanta per omnes gentes nationesque linguæ diversitate, hic mihi omnium hominum communis sermo videatur," etc.

admire, numerate, confess, repent, fear, . . . and what not? we find a variety and multiplication which might well be the envy of the tongue";[59] and further on, in the same chapter, he expressly names cheiromancy, giving a few of its indications and adding:—"I call you yourself to witness whether with this science a man may not pass with reputation and favour in every company."[60] In France, besides Montaigne, Honoré de Balzac has given great attention to the subject. Théophile Gautier calls special attention to the fact in his work, "Honoré de Balzac" (Paris, 1860, p. 165), and certainly we find long passages on the science in Balzac's "Comédie Humaine,"[61] to which I beg particularly to refer the reader, and in the course of which he remarks:—"To foretell to a man the events of his life, from the aspects of his hand, is not a thing

De Balzac.

His arguments for Cheirosophy.

[59] "Essais de Montaigne, suivis de sa corréspondance et de la servitude volontaire d'Estienne de la Boëtie," etc. (Paris, 1854); "Apologie de Raimond Sebond," vol. ii., p. 282, book ii., ch. xii.:— " Quoy des mains ? nous requerons, nous promettons, appelons, congedions, menaçeons, prions, supplions, nions, refusons, interrogeons, admirons, nombrons, confessons, repentons, craignons, vergoignons, doubtons, intruisons, commandons, incitons, encourageons, jurons, tesmoignons, accusons, condamnons, absolvons, injurions, mesprisons, desfions, despitons, flattons, applaudissons, benissons, humilions, mocquons, reconcilions, recommendons, exaltons, festoyons, resjouïssons, complaignons, attristons, desconfortons, desesperons, estonnons, escrions, taisons, et quoy non ? d'une variation et multiplication, à l'envy de la langue."

[60] "Il ne fault sçavoir que le lieu de Mars loge au milieu du triangle de la main, celuy de Venus au poulce, et de Mercure au petit doigt ; et que quand la mensale coupe le tubercle de l'enseigneur, c'est signe de cruauté ; quand elle fault soubs le mitoyen et que la moyenne naturelle faict un angle avecques la vitale soubs mesme endroict, que c'est signe d'une mort miserable, etc. Je vous apelle vous mesme à tesmoing, si avecques cette science un homme ne peult passer, avec reputation et faveur, parmy toutes compaignies."—OP. CIT., ch. xii., p. 470, vol. ii.

[61] Alphonse Pagès, " Pensées de Balzac extraites de la Comédie Humaine " (Paris, 1866), livre v., La Société ; cap. vi., " Sciences Occultes."

AN INTRODUCTORY ARGUMENT. 63

more strange for him who has the qualities of a seer, than it is to tell a soldier that he will fight, a cobbler that he will make shoes, or an agriculturist that he will dress and work the soil." And again he says :—" Many sciences have issued from the occult ones, and their illustrious discoverers have made only one mistake, which is—that they try to reduce to a system isolated examples, of which the creative cause has not yet been able to be analysed." The passages are too long to transcribe [even as notes] in this place; but I warmly recommend them to my readers. It is not necessary to multiply authorities. What I have given above will carry their own weight; and now, before recapitulating the claims of the science to our consideration, I should like to say a few words on the astrological aspect of the subject.

¶ 61. Astrologic Cheiromancy

As to the nomenclature which has been adopted for the mounts, I have explained its object in another place [*vide* ¶ 371]; what I wish particularly to notice here are the astrologic explanations which have been, by many writers, advanced for the establishment of the science, with a few words on their arguments in support of the hypothesis, and my own view of the matter.

¶ 62. Influence of the planets upon the earth.

Much has been said [though little has been definitely known] for and against the influence of the sun, moon, and planets upon the earth and the people inhabiting it. The question seems to turn upon the existence of a connecting link which joins us to them, a connecting fluid, the principal function of which is the transmission of their light and heat to our globe :

Æther.

and this connecting fluid is what is known as *æther*. The problem of the existence in all space of a fluid called æther or its non-existence was first enunciated by Zuglichen van Huyghens [nat. 1619, ob. 1693], who was the first to propound the undulatory as

Van Huyghens and Newton.

opposed to the molecular theory of light, which latter was the then generally accepted theory of Newton. Van Huyghens took from the analogy of sound in air and waves in water the idea of the existence *in all space* of a highly elastic [quasi-solid] fluid, provisionally termed *æther*, and started the now well-known **Undulatory theory of light.** and accepted hypothesis that light consists of the propagation of waves in this fluid. The hypothesis also requires that the vibrating medium should possess properties more nearly allied to those of an elastic solid than those of a vapour or gas.

¶ 63. The theory developed. Young. These two theories [the undulatory and the molecular] were pretty evenly balanced in scientific and popular estimation, until in 1802 Dr. Young, by his discovery of the laws of the interference of light, turned the scale in favour of the undulatory as against **Fresnel.** the molecular theory. Twelve years later, Fresnel [between 1814 and 1819], in ignorance, it is said, of the labours of Young, " demonstrated to his countrymen the error of the Newtonian theory of the propagation of light by the emission of material particles, and ably advocated the undulatory hypothesis."[62] This is, I think, all that need be said in proof of a connecting quasi-solid matter or fluid beyond the atmosphere in which the entire solar system floats,—a fluid sufficiently ponderable to resist the passage of comets, and consequently of transmitting to us the influences of the moon and stars, influences sufficiently demonstrated by the phenomena of tides, without going into their influence upon certain persons in various conditions of mind and body.

[62] A full account of the discoveries of Fresnel in this connection may be found in the "Œuvres Complètes d' Augustin Fresnel, publiées par les soins du Ministre de l'Instruction Publique" (Paris, 1866), at p. 247 of vol. i. of which will be found No. XIV., "Mémoire sur la diffraction de la lumière, couronné par l'Académie des Sciences, 1819," in which Mémoire his studies and their results on this point are fully set forth.

AN INTRODUCTORY ARGUMENT.   65

"If therefore," says Daubenay, in one of his speeches before the Royal Society, "the direction of a rod of steel hung a few feet from the earth can, as has been proved by Colonel Sabine, be influenced by the position of a body like the moon situate 200,000 miles [mean distance 238,750 miles] from our planet, who can accuse of extravagance the belief of the ancient astrologers in the influence of planets on the human system?" And if the heavenly bodies can act through two hundred and odd thousand miles of that connecting fluid or æther, on an inanimate object like a piece of steel, why should they not act, and act so strongly as to influence our whole lives, on so sensitive, so impressionable a substance as that imponderable nerve fluid [vide ¶ 49] which is our life, our sense, our very soul? ¶ 64. Daubenay. Colonel Sabine's experiments.

But we are going further than I intended; the astrologers who laid down the hard-and-fast rule that our existences are directed by the states and positions of the planets at the time of our birth, seem to have passed over the influences of parentage as being immaterial, and take into no account the physical effects of the mental and physiological conditions of our progenitors. And again, they err who say that all things hereditary are inevitable for the reason that we do not choose our own parents, because in a manner they are chosen for us; that is to say, our parentage is pre-ordained, for it is the inevitable and continual march of events which gives us our ancestors, and, as a matter of fact, our parents, the time of our birth, and many of the other influences of our lives are merely the results of the natural sequence of certain already established facts, to the examination of which the ancient astrologers devoted their lives with such assiduity. Turn, if this seems incomprehensible to you, to the opening lines of Dugald Stewart's immortal work:[63]—"All the different ¶ 65. Astrology Choice of parents Sequence of events. Dugald Stewart

₆₃ DUGALD STEWART, "Outlines of Moral Philosophy," 6th ed., 1837.

kinds of philosophical inquiry, and all that practical knowledge which guides our conduct in life, presuppose such an established order in the succession of events as enables us to form conjectures concerning the future from the observation of the past." Philosophy, therefore, aims at ascertaining the established conjunctions which, in their turn, establish the order of the universe; the result of possible combinations of future events become known to us by means of those *artificial combinations of present circumstances* which are known to us by the name of *experiments;* and as Dugald Stewart, in the same work, has remarked :—" Knowledge of the laws of nature is to be attained *only* by experiment, for there is no *actual* connection between two events which enable us to form an *à priori* reasoning.[64] [*vide* ¶ 89 and 90.]

¶ 66.
Herbert Spencer.

The laws of natural causation.

Herbert Spencer, in his "Study of Sociology" [London, 1873], deals very ably and interestingly with this point in his chapter [II.], entitled, " Is there a Social Science," at the end of which he says :—" In brief, then, the alternative positions are these. On the one hand, if there is no natural causation throughout the actions of incorporated humanity, government and legislation are absurd. Acts of Parliament may, as well as not, be made to depend on the drawing of lots or the tossing of a coin ; or rather, there may as well be none at all ; social sequences having no ascertainable order, no effect can be counted upon ; everything is chaotic. On the other hand, *if there is natural causation,* then the combination of forces by which

" Balzac says on this point :—" Que certains êtres aient le pouvoir d'apercevoir les faits à venir dans le germe des causes, comme le grand inventeur aperçoit une industrie, une science, dans un effet naturel inaperçu du vulgaire ; ce n'est plus une de ces violentes exceptions qui font rumeur, c'est l'effet d'une faculté *reconnue*, et qui serait en quelque sorte le sonnambulisme de l'esprit. Si donc cette proposition, sur laquelle reposent les différentes manières de déchiffrer l'avenir, semble absurde, le fait est là."

AN INTRODUCTORY ARGUMENT.

every combination of effects is produced, produces that combination of effects in conformity with the laws of the forces. And if so, *it behoves us to use all diligence in ascertaining what the forces are, what are their laws, and what are the ways in which they co-operate.*"

¶ 67. Natal influences

Thus, therefore, it is not in any way absurd to study the atmospheric, the meteorologic, or, if you will have it so, the astrologic conditions under which a man is born, and under which his parents have lived, in making a probable forecast of the tendencies and even of the events which will signalize his life; but in all, and through all, we must bear in mind the thought which Desbarrolles has embodied in this striking sentence:

Desbarrolles.

—" The influence of the planets is incontestable; but what is still more incontestable is the universal and all-powerful action of a Being supremely pre-eminent, who rules and governs the stars, the heavens, the visible and the invisible worlds, unlimited space, and the immensity of the universe. This Being, whom our dazzled reason cannot conceive, this Being whom our reason adores, but to whom it dares not give a name, has been named by mortals—God." [65]

¶ 68. Astral lines.

The astrologic cheiromant lays down as an axiom that the lines and formations which exist in a hand at the moment of birth are purely astral, are produced by the influences of planets which have been at work up to that moment, and that it is the action of the brain-development which modifies them afterwards. The physiological cheiromant, on the other hand [and

---

[65] ADRIEN DESBARROLLES : " Les influences des astres sont incontestables; mais ce qui est plus incontestable encore, c'est l'action universelle et tout-puissante d'un être éminemment supérieur, qui régit les astres, les cieux, les mondes visibles, les mondes invisibles, les espaces sans bornes, l'immensité! Cet être, que notre raison éblouie ne peut concevoir, cet être qu'elle adore et auquel elle n'ose donner un nom, les hommes l'ont appelé Dieu."—" Les Mystères de la Main, révélés et expliqués ' (Paris, 1859).

to this opinion I incline myself], considers rather that the *tendencies of a man's nature* are the result of his ante-natal and ancestral circumstances, that it is these tendencies that mould the formations of his hands, and that the events and characteristics of his life may be explained and foretold by a careful study of those *causes* [*i.e.*, those tendencies] based upon *experiences* which, in these cases, do duty for experiments.

¶ 69.
The ultimate and the proximate cause.

The whole question, therefore, of the astral influences, with regard to the science of Cheirosophy, resolves itself into a consideration of the ultimate and proximate cause, and my view of the case is this: Let us firmly establish and recognise the proximate cause [*i.e.*, the physical conditions and ante-natal circumstances of man], before we begin to search for the ultimate cause, [*i.e.*, the forces and influences which have caused those physical conditions and ante-natal circumstances]; let us fully understand our physical position, before we soar away into speculations [for mere speculations they must be] as to our psychical position: to my mind, the main obstacle to the proper elucidation of such problems as these, is nothing more or less than the fact that investigators are ever prone to attempt to explain psychical problems [which depend naturally upon physical ones], before they have thoroughly examined the physical conditions which must lead to that explanation. This, then, is my position as regards astrology in connection with Cheirosophy: let us clearly understand, what is clearly capable of comprehension, before we speculate on ultimate causes, concerning which we can, in our present condition, and with our present means of information, know little or nothing. Let us wait and study patiently:

Alexandre Dumas fils

"La Chiromancie," says Alexandre Dumas fils, "sera un jour la grammaire de l'organisation humaine."

AN INTRODUCTORY ARGUMENT. 69

¶ 70. Final argument

We reach now the final, [and perhaps the most important] point of this argument. I wish to conclude as briefly as possible [for my arguments have unconsciously been extended] with a few of the considerations connected with the actual science of Cheirosophy, which entitle it to the most careful and universal study; which fully demonstrate the value of its indications, and which establish its claims to rank as an exact science : my principal difficulty will be to confine myself to a statement of the more important points, so as to avoid losing the point of my argument, in the diffuseness of its treatment.

¶ 71. Primary propositions.

And, firstly, let me deal with that branch of argument, which ridicules the idea of seeking in the human hand the indications of the human character. [*Vide post* ¶¶ 78ª and 96.] There are but few people who will disagree with me at the outset, if I lay down the proposition, that no two characters are absolutely identical, any more than two faces, or two methods and manners of speech are so. Bearing this in mind, it is interesting and pertinent to note that no pair of hands are exactly alike : indeed, we may go further and say that it is impossible to find two pairs of hands which do not exhibit very striking and plainly marked differences, both as regards their forms and shapes, and as regards the markings found therein.[66]

¶ 72. Infinite differences. The Chinese.

The Chinese have a system of divination by the examination of the impression left by the ball of the thumb upon a piece of soft wax, or from the oval figure which may be traced from it upon a piece of paper, using the thumb as a die, and daubing it with ink. It is a matter of common knowledge that the spiral and convoluted figures, produced by any

---

[66] "Il y a autant de diversité et de dissemblance entre les formes des mains, qu'il y en a entre les Physionomies."—J. C. LAVATER, "L'Art de connaître les hommes par la physionomie" (Paris, 1806), vol. iii., p. 1.

unlimited number of thumbs, will all present some difference one from another. It is true that, as Aristotle says: "In some animals there is a mutual resemblance in all their parts; as, the eye of any one man resembles [in construction] the eye of every other man; for in individuals of the same species, each part resembles its correspondent part, as much as the whole resembles the whole;"[67] and so it is of the hand, *i.e.*, every hand resembles its fellows in that [barring the cases of deformities] it has a thumb, four fingers, a palm, and so on; but the composition and formation of those parts of the member, differ invariably, and *ad infinitum;* and here we have two most pertinent and leading facts, that, like as all dispositions are different, so all hands are different; and who shall accuse of absurdity the proposition, that these two facts may bear a very close relationship to one another? And, as we have seen that the arrangements of veins and nerves in a hand vary indefinitely [*vide* ¶ 38], may not the constitutional [the constructional] variations thereby indicated be the first step towards the explanation of those differences of character, which trace themselves at the point where the actual mechanical arrangement shows the greatest variation, and that is—in the hand?

¶ 73.
Use indicated by aspects.

From the paw of a dog, you can tell what sort of chase he will be most useful for; from the shape of a horse's hoof, you can tell his breed, and the kind of work to which he is best adapted. Why, therefore, should we not be able to tell from the hand of man what are his principal occupations, and the consequent tendencies of his nature? To carry this argument a

---

[67] ARISTOTLE, ΠΕΡΙ ΤΑ ΖΩΑ ἹΣΤΟΡΙΟΝ, Βιβλ. ἀ., Κεφ.. ἀ:—
" Ἔχει δὲ τῶν ζῴων ἔνια μὲν πάντα τὰ μόρια τ'αὐτὰ ἀλλήλοις, ἔνια δ' ἕτερα. Τ'αὐτὰ δὲ τὰ μὲν εἴδει τῶν μορίων ἐστίν, οἷον ἀνθρώπου ῥὶς καὶ ὀφθαλμὸς ἀνθρώπου ῥινὶ καὶ ὀφθαλμῷ, καὶ σαρκὶ σὰρξ, καὶ ὀστῷ ὀστοῦν· ὁμοίως γὰρ ὥσπερ τὸ ὅλον ἔχει πρὸς τὸ ὅλον, καὶ τῶν μορίων ἔνει ἕκαστον πρὸς ἕκαστον.

little further, let us remember Sir Charles Bell's most interesting disquisition on the identification of bones; and his account of how from a chance fragment of bone the zoologist can re-construct an entire animal, beginning by its formation and consistency [whereon *vide* ¶ 36], which betray the habits of the animal; the shapes of the jointures, and processes, whereby we learn the use of the limb of which it formed part; and thence onward to the manner of obtaining nutriment and its nature, from which data the entire habits of the animal may be accurately ascertained. Why, therefore, should we not be able, as it were, to "re-construct" the MAN, his habits, and his characteristics, from the examination of the HAND, the prime agent of his character, and of his will? As D'Arpentigny says: "What shall we think of the Indian law, which obliges a son to follow the profession of his father: is it not evident that the legislature would have done better in ordaining that men whose *hands* were similar, that is to say, whose *instruments* were the same, should devote themselves to the same labours?" *[marginalia: Sir C. Bell. The identification of bones. D'Arpentigny]*

Let us carry this illustration yet further: a farmer sees furrows in a field; if he is an intelligent man and knows his business, he can tell not only that a plough caused them, but he can tell what kind of plough and how it was used. In like manner can the Cheirosophist interpret the meanings, with their causes, of the lines, the furrows which are traced upon the human hand. By use an organ becomes developed, by nonuse it becomes feeble and enervated [*vide* again ¶ 36]. This is a truism, but it is germane to our subject, for from the development of the hand we find out what its use has been. *[marginalia: ¶ 74. Illustration.]*

Now it is constantly flung at the student of Cheirosophy that the lines and formations are merely caused mechanically by the *folding* of the hands, and by the *[marginalia: ¶ 75. The "folding" argument.]*

use which their owners make of them, and that the more the hands are used, the greater the number of lines which will be apparent in them; but this is *Contradictions.* directly contradicted by the *fact* that the hands of the labouring classes are, with the exception of the principal lines, almost devoid of any markings whatsoever, whereas those people whose hands have hardly any rough work to do at all [especially those of women of leisure, who *never* work with their hands] are nearly always covered with a perfect network of lines. Not using their hands in active employments, the brain develops an increased activity of the hands, whereby the lines become traced, and has more time to cause its impressions to become written on the most sensitive surface that it can find, and with which it is the most intimately connected, and that is in the HAND. I have often wondered how people who cling to the " folding " argument, would account for two facts which they seem utterly to ignore. Firstly, what pro-
Lines at points duces the lines which we find at points where no folding
of no flexion. ever by any possibility can take place in a direction which would produce the lines we find at these points? Lines so placed, not being produced by any mechanical cause which is immediately apparent to these arguers, must of necessity mean something even to them, for it is one of the first rules of Ethics, that God and Nature do nothing without a purpose. "*Deus et Natura nihil efficiunt frustra.*" And secondly, how do they account for the fact that we find lines traced deeply and clearly [more so very often than at a later *Lines in the* age] in the hands of children at the moment of their
hands of infants. birth? I have seen lines in the hands of a child only a few hours old, which have entirely altered within the first few months or years of its life; this is a circumstance which is strenuously put forward by the astrological Cheiromants in support of their views, but which, to my mind, is accounted

for in the manner I have described in another place [¶ 65]. Mr. Ebule-Evans has said on this point in the pages of *St. Paul's Magazine* [vol. xiii., p. 332], "True, the *proximate* cause of these lines is the various motions of the hands; but what gave your hands these particular varieties of motion? Is it not as easy [and much more natural] for fate to guide the motions of your hand, so that its lines, rightly interpreted, shall exactly represent your career, as it would be to stamp these lines in an arbitrary position on your chest, as men impress the image of a cow on butter? And yet were the latter done, science would have an objection the less to urge!" Ebule-Evans.

Every type of hand [*vide* ¶¶ 155 and 164] has certain characteristics which may be *altered* or *modified* by forcing it to a labour entirely opposed to the inclination or talent of which it is the indication, but no work can obliterate, far less *alter*, the original shape and formations of a hand. It is only natural, and indeed it is inevitable that a hand should betray the occupation of a body or mind, more than anything else; this obvious axiom seems to have struck Honoré de Balzac, when he makes the remarks quoted in ¶ 60, p. 62. ¶ 76. Modification of types.

There are some who have not the courage utterly to deny that Cheirosophy is entitled to rank as an exact science, and who compromise by saying that it is the "shadow of a truth;" I would submit in reply that from its shadow the form of any object may be ascertained and determined in its entirety, and that the phrase is bad from a comminatory point of view, because it places Cheirosophy in the position of an interpreting diagram, which is as it were the reflection of Nature. ¶ 77. Half-way sceptics.

Again, many of the "half-way" sceptics will say, "We can believe in Phrenology and Physiognomy, but not in this." A moment's reflection will cause us to remember and appreciate the circumstance that, as ¶ 78. Phrenology and Physiognomy. Their superiority over Cheirosophy discussed regards physiognomy, the expression of the face can be varied at will, and by constant attention the features may be permanently set into an entirely misleading expression; but the hand, no; the hand cannot be altered, be the effort to do so ever so strong. "For the hand has its physiognomy like the face, only, as this particular physiognomy reflects only the immovable basis of the intelligence, it has all the immobility of a material symbol. Mirror of the sensations of the soul, of the heart, of the senses, and of the spirit,—the physiognomy of the face has all the charms of variety; but, as to a certain extent, it may be dominated by the will; nothing can guarantee to us the truth of its revelations, whilst the hand preserves invariably the same expression, whatever it may be, of our natural bent;" [68] and Balzac, in the work I have recently quoted, has been struck by the same thing; for he says:—"We acquire the faculty of imposing silence upon our lips, upon our eyes, upon our eyebrows, and upon our forehead; the hand does *not* dissemble, and no feature is more expressive than the hand.[69] The hand has a thousand ways of being dry, moist, burning, icy, soft, hard, unctuous, it palpitates, it perspires, it hardens, it softens. In fact, it presents an inexplicable phenomenon that one is tempted to name the incarna-

[68] D'Arpentigny, "La Science de la Main" (Paris, 1865), p. 94.

[69] The physiognomist Lavater, in the work I have already quoted, Note [66], p. 69, continues the passage I have there noticed by saying:—"La main est ... un object de la physionomie, un objet d'autant plus significatif, et d'autant plus frappant parce-que la main ne peut pas dissimuler, et que sa mobilité le trahit à chaque instant. Je dis qu'elle ne peut pas dissimuler, car l'hypocrite le plus raffiné, le fourbe le plus exercé, ne saurait altérer ni la forme, ni les contours, ni les proportions, ni les muscles de sa main, ou seulement d'une section de sa main; il ne saurait la soustraire aux yeux de l'observateur qu'en la cachant tout-à-fait."

tion of thought. . . . In all ages sorcerers have tried to read our destiny in the lines which are in no way fantastic or meaningless, and which correspond with the principles of the life and of the character." And we may here call attention to the point, and I have noticed it in another place [¶ 108] that the *Cheirognomy*, the shapes of a hand, are very frequently hereditary, whilst the lines of the palm become traced by the other and more exterior influences to which I have already alluded. Then, as regards the superiority of phrenology, here again I beg leave to differ; by calling attention to the points that in that science it is very easy to mistake the position of a bump, or to ascribe to it an erroneous signification by reason of its displacement, and that by reason of the growth of the hair it is difficult to put into practice, and indeed, cannot be exercised without the consent of the subject, whereas in Cheirosophy the positions of the formations are much more clearly defined, and every displacement of a line or mount carries with it its own particular signification, whilst it can be put into operation without the consent or even the knowledge of the subject of your scrutiny.

Heredity of Cheirognomy.

"It is the word 'absurd'" (says Balzac), "which condemned steam, which condemns to-day aërial navigation, which condemned the inventions of gunpowder, of printing, of spectacles, of engraving, and the more recent discovery of photography. . . . Very well then, if God has traced for certain clear-sighted beings the destiny and character of every man in his physiognomy [taking this word to mean the whole expression of the body], why should not the hand resume in itself the whole of the science of physiognomy, seeing that the hand represents human action in its entirety, and its only mode of manifesting itself? And thus we attain to Cheiromancy." ["Comédie Humaine," book V., cap. vi.]

¶ 78ª. *Ridicule. De Balzac.*

¶ 79.
Value of sceptic arguments.

It must not, however, be thought that I object, in the slightest degree, to the scepticism with which the science is frequently received; on the contrary, the sceptic acts towards Cheirosophy in the relation that darkness bears to light, *i.e.*, it brings it into prominence, and, indeed, is the main evidence of its existence; for, as without shadow, light could not be proved to exist, so without scepticism the truths of Cheirosophy would be lost in the unquestioned presence of their evident and eminent reality.

¶ 80.
Evils of impatience in argument.

In all my arguments on this science I have strenuously endeavoured to avoid that irritation and impatience which is too often the inseparable concomitant of argument. "Along with the irrational hope so conspicuously shown by every party having a new project for the furtherance of human welfare, there habitually goes this irrational irritation in the presence of stern truths which negative sanguine anticipations. Be it . . . some plan for reforming men by teaching; . . . anything like calm consideration of probabilities as estimated from experience is excluded by this eagerness for an immediate result; and, instead of submission to the necessities of things, there comes vexation, felt, if not expressed, against them, or against those who point them out, or against both."[70] I have laboured to avoid this fault by courting adverse criticism with a thankful appreciation of its value.

¶ 81.
Wickedness of the science.

Some persons there are who actually look upon the science as something wicked and uncanny, averring that it is not to be permitted to presumptuous man to read the secrets of the Most High.

¶ 82.
Degeneration of the science.

That such a line of argument should be taken up by even the most narrow-minded sectarians, can only result from the evil repute into which the science at one time undoubtedly fell,—a state of things which

---

[70] HERBERT SPENCER, "Study of Sociology" (London, 1873), chap. vii., "Emotional Subjective Difficulties."

called forth from Ferrand the expostulation, I might almost say the "Cheiromantiad," with which I have headed the Introduction to a former work :[72]—"This art of Chiromancy hath been so strangely infected with superstition, deceit, cheating, and (if I durst say so), with magic also, that the Canonists, and of late years Pope Sixtus Quintus, have been constrained utterly to condemn it; so that now no man professeth publicly this cheating art, but theeves, rogues, and beggarly rascals, which are now everywhere knowne by the name of Bohemians, Egyptians, and Caramaras, who have arrived in Europe since the year 1417, such as G. Dupreau, Albert Krantz, and Polydore Vergil."[73]

<small>J. Ferrand.</small>

To those who would say, in the words of the Hierophant in the *Zauberflöte* [Act I., Sc. xvi.] :—

<small>¶ 83.
Legitimacy of the inquiry.</small>

"Wo willst du kühner Fremdling hin?
Was suchst du hier im Heiligthum?"

I answer that the Book of Nature is open to all men to read, but that Nature imposes the necessity of assiduous study, before she will surrender the secrets she has veiled, with a not impenetrable cloud of obscurity. If this is not so, why has she marked the past, the present, and the future on the hands of man, who, if he will devote himself to the study, may read them? Why has she marked indelibly and unalterably upon the hands of the hypocrite his real character, which, written on his head or face, he may conceal by the growth of his hair, or the distortion of his features?

Other theological enemies of the science aver,—by

<small>¶ 84.
Atheism.</small>

---

[71] ED. HERON-ALLEN, "Codex Chiromantiæ" (Odd Volumes Opuscula : No. VII.), London, 1883.

[72] JACQUES FERRAND, "De la Maladie d'Amour, ou Melancholie Erotique" (Paris 1623), ch. xxii., p. 134. *English translation :* "Erotomania; or, Love Melancholy" (London, 1640), p. 173.

what mode of reasoning I know not,—that Cheirosophy argues a disbelief in the existence of a Deity, and of a future state; on the contrary, it constantly brings before the student the evidences of an all-powerful Agency, and constantly directs his thoughts, both to the immediate future, which concerns us all so nearly, and to the ultimate future, which should concern us still more.

Desbarrolles As Desbarrolles has said, at p. 484 of his *magnum opus*:[72a]—"They wish to impede the progress of Cheiromancy, under the pretext that it is wrong to go beyond the limits of natural knowledge; but do not *spectacles*, which restore to the sight the vigour of youth, trespass beyond the limits of nature? must they, on that account, be proscribed? and microscopes, which make visible the invisible? and telescopes, which reveal the nature of the heavens?" [*vide* ¶ [78a].]

¶ 85.
Painful duties of the science. I do not deny that there is a painful side to the science: that the knowledge which we obtain is often terrible and saddening, betraying the faults and the misfortunes of our friends, as well as our own, and often dissipating our most fondly-cherished illusions;

Parallel to astronomy. but who dares to deny the inestimable value of the science? The astronomer, in the observatory, predicts a storm, the means whereby the sailor's life may be endangered [not his certain death], and the sailor does not embark; a few days or hours later the storm supervenes, and the sailor's life is not wasted. So the Cheirosophist predicts a blow by the observation of the tendencies which will bring about a misfortune; the subject takes steps to escape it, and the blow falls harmless. Aristotle spoke truly when he said, "*Homo sapiens dominabitur astris.*" The astronomer warns by the promptings of astrology, the influence of the heavenly bodies upon the earth; the astrological Cheiromant will tell us that he also warns

---

[72a] ADRIEN DESBARROLLES, "Mystères de la Main." Révélations complètes. Suite et Fin (Paris, 1879).

AN INTRODUCTORY ARGUMENT. 79

by astrology, as the basis of the science [*vide* ¶ 68]; and it is Cheirosophy which, thus based upon the highest natural influences, finds out our natural tendencies and the influences under which we principally are, and enables us to see in ourselves and others the rules by which the life, the actions, and the destiny are governed. And Cheirosophy, thus based upon astrology [if you will have it so], physiology, and ethics, gives to youth the experience and foresight of age, endows all men, who will study it, with that prescience which, under the name of intuitive faculty, is the cherished possession of so few, and enunciates and solves the problem of " Know thyself."

<small>Effects of Cheirosophy.</small>

ΓΝΩΘΙ ΣΕΑΥΤΟΝ.

Again, it reveals the natural aptitude of the young mind, and points out the walk in life to which it is most adapted [73]; it points out the obstacles which beset the life, and how to avoid them; it is the pilot, which will, if properly consulted and obeyed, take us through the shoals of human tendencies; and thus the science once proscribed, but now divested of its deceit, its mystery, and its charlatanry, shines forth with the radiance of a pure science, courting the daylight of scientific investigation, the tests of adverse criticism, time, and fanatic opposition, and the concentrated attention of all *reasoning* beings.

<small>¶ 86. Indications in youth.</small>

The writer in *St. Paul's Magazine*,[74] whom I have before quoted, remarks in connection with this

<small>¶ 87. Ebule-Evans' arguments.</small>

---

[73] "Mais combien est petit le nombre des jeunes gens auxquels il arrive d'être divinés assez à temps pour être bien dirigés ! et combien aussi est petit le nombre des précepteurs qui consentent, abdiquant tout système exclusif, à adopter un système à part pour chaque génie à part. Ce ne serait pas trop attendre de la sollicitude d'un père, mais cet effort [et il faudrait le ranger parmi les plus généreux] sera toujours au-dessus de la sollicitude vénale d'un étranger."—D'ARPENTIGNY.

[74] A. EBULE-EVANS, *St. Paul's Magazine*, vol. xiii., p. 332, "Chiromancy."

science:—"It is as unscientific to gauge the credibility of moral phenomena by physical tests as it would be to attempt to demonstrate physical phenomena by arguments drawn from the region of moral speculation . . . the *logical* lesson to be learnt from this is, Study them more closely, and endeavour to get at their explanation; the *scientific* conclusion actually drawn is, We cannot explain them, *ergo* they are all humbug! I venture to say that this conclusion will satisfy those alone who have never studied the history of science, and who are, therefore, not aware that every addition to our knowledge has been made in the teeth of scientific opposition." It seems to me to be a very easily conceivable thing to realize that all these indications are brought about by what we may call, for the sake of definition, "Currents of Impression," by the agency of the highly sensitive *nerve fluid* to which I have referred at length, which comes directly from the brain, and is amassed in the greatest quantity in the palm of the hand. Now, this constant passing action of a fluid endued with this great power must of necessity act upon the delicate tissues in which it works, just as the continued drop of water, or the constantly recurring footstep wears away the hardest granite; and, knowing that this powerful agency is incessantly at work in the exquisitely sensitive palm of the hand, which is [as my whole argument has continued to show] the most important auxiliary of the sense of touch, is it rational to say that the lines there found are the result of "accident," and that Omnipotent Nature [which "does nothing in vain"] does not direct or control the clearly designated features of her most sensitive organ?

Currents of impression.

¶ 88.
Prediction.

It is often considered the leading argument on the charlatanry of the science, that the Cheirosophist claims to predict a future malady, and even the time of the death of the subject whose hand he holds in his own; but a moment's rational thought will bring the conviction that this is no charlatanry or empty arrogation of power. It must be acknowledged that in every system there must lurk the germ of the disease which will ultimately supervene, and which may prove fatal; at a greater or less distance of time possibly, but still there it exists, and, as it is destined to play so important a *rôle* in the history of one's life, the Omniscient Soul must be aware of it, though its knowledge may be unconscious, or rather, I would prefer to say, supra-conscious; and imperceptibly this knowledge becomes more and more strongly developed till the malady supervenes, and passes away, leaving its trace behind, *or*, carrying us to that bourne whence no traveller returns: *and*, as of a line which is destined to appear, the root must necessarily exist already in the hand, it is the study of the Cheirosophist to discover that root, and predict the time of its full development and its probable effects. Who shall deny the presumption that the germs of a future disease produce certain effects upon the nerve fluid which direct the manner in which it works in the arrangement of the tactile corpuscles [*vide* ¶ 47] to produce the lines in our hands? [Germs of diseases.] ["Line-roots.]

That the tendencies of our natures combined with other circumstances of our lives, shown by the examination of our hands, are such as will lead, *if unchecked*, to certain specified results; that these things *are*, there is no possibility of denial, and it is the duty of the Cheiromant to point out those ends to which he sees an existence tending; but be it most distinctly understood [and on this I cannot lay too much stress] that the science of Cheirosophy never pretends to say, "What is written *shall be*," only this, that it possesses the power of warning us of events which, unless controlled, will come to pass, to the end that, the warning being given and accepted, the subject [¶ 89. Tendencies which lead to results.] [Non-fatalism of the science.]

under examination may so bridle himself as to obviate the results [if bad] which will in all probability supervene if permitted so to do.[75] Thus the Cheirosophist can say with Demosthenes :—"You need not pry into the future; but assure yourselves it will be disastrous unless you attend to your duty, and are willing to act as becomes you."[76]

Demosthenes on the future.

¶ 90.
As to the future.

The whole question of prediction turns, as will be seen, on this question of the doctrine which lays down the proposition that the *future* is the result of the *present*, and that the result of present circumstances [*i.e.*, the future] may be foreseen by minds specially trained to the minute analysis that this requires. It turns also on the question whether the future *exists* or not, that is to say, does "the middle of next week" exist *at this moment*, or does it only come into actual existence as the time comes round; if it does *not* exist [*i.e.*, if we could be suddenly projected "into the middle of next week," and were to find chaos, *nothing* (if such a state is remotely conceivable)], then it is simply manufactured by present events, and is at

---

[75] Balzac on this point says :—"Remarquez que prédire les gros événements de l'avenir n'est pas pour le voyant un tour de force plus extraordinaire que celui de deviner le passé. Le passé, l'avenir sont également impossibles à savoir, dans le système des incrédules. Si les événements accomplis ont laissé des traces, il est vraisemblable d'imaginer *que les événements à venir ont leurs racines* [*v. sup.*]. Dès qu'un diseur de bonne aventure vous explique minutieusement les faits connus de vous seul dans votre vie antérieure, il peut vous dire les événements que produiront les causes existantes. Dans le monde naturel les mêmes effets s'y doivent retrouver avec les différences propres à leurs divers milieux."

[76] "Οὐ γὰρ ἄττα ποτ' ἔσται δεῖ σκοπεῖν, ἀλλ' ὅτι φαῦλ' ἂν μὴ προσέχητε, τοῖς πράγμασι, τὸν νοῦν καὶ τὰ προσήκοντα ποιεῖν ἐθέλητ', εὖ εἰδέναι."—ΔΗΜΟΣΘΕΝΟΥΣ ΚΑΤΑ ΦΙΛΙΠΠΟΥ. Α΄. And again : Εἰσὶ τοίνυν τινὲς οἳ τότ' ἐξελέγχειν τὸν παριόντα οἴονται ἐπειδὰν ἐρωτήσωσι, τί οὖν χρὴ ποιεῖν; οἷς ἐγὼ μὲν τὸ δικαιότατον καὶ ἀληθέστατον τοῦτο ἀποκρινοῦμαι ταῦτα μὴ ποιεῖν ἃ νυνὶ ποιεῖτε.— ΠΕΡΙ ΤΩΝ ΕΝ ΧΕΡΣΟΝΗΣΩ (38).

AN INTRODUCTORY ARGUMENT. 83

the present moment in the process of manufacture, and the task of the seer is greatly simplified, for he can so direct present circumstances as to evolve a certain future; but if, as is more likely the case, the future is pre-existent as the flight of swallows, the variations of the barometer, and the premonitory symptoms of disease would indicate, the task of the cheirosophist becomes more complicated, for whatever shall be, *now is*; and it is his apparently marvellous task to raise a powerful presumption of its exact nature. "Our thoughts," says Professor Owen, "are free to soar as far as any legitimate analogy may seem to guide them rightly in the boundless ocean of unknown truth!"[77] *Owen.*

It is in dealing with future events and their traditional signs that the Cheirosophist finds the greatest difficulty and room for the greatest doubt. Thus, in the case of sudden deaths and unforeseen calamities, we cannot satisfactorily account for the signs which predict them; but it is a fact that certain traditional signs are accepted by the schools of Cheirosophy to indicate certain unforeseen occurrences, and until they shall fail to predict correctly, we must accept them and retain them, but use them and cite them warily and discreetly. Cardan remarked that of forty-five persons to whom Barthelemy Cocles had predicted a sudden death, only two failed to fulfil his prediction.[78] M. Desbarrolles, in the introduction to the 15th edition of his work already cited [note [65], p. 67], says on this point:—" I warn the reader that of the ancient Cheiromancy, and particularly of the prophetic nonsense ¶ 91. Traditional signs.

Cardan. Cocles.

Desbarrolles

---

[77] Professor Sir RICHARD OWEN, Op. cit., p. 83.
[78] BARTHOLOMÆUS COCLES, "Chyromantie ac Physiognomie Anastasis," etc., 1503, fol. Bartholomæi Coclitis "Physiognomiæ et Chiromantiæ compendium" (Argentorati) 1533 [1534, 1554, and 1555]. "Le Compendion et brief enseignment de Physiognomie et Chiromancie" de B. Cocles (Paris, 1550).

of the sixteenth century, I have only retained certain signs which, repeated by all authorities, deserved some consideration; I have found them to be correct, and I have adopted them; but I must add that since the first edition of my book [now some ten years since] I have not found, in the innumerable applications I have made of them, a single instance in which I could recognize their exactness." And, in his larger and later work, he continues, "It is good to seek for the meanings of popular traditions; there is always evidence of a *something* which has preserved the idea from oblivion. Good sense is often hidden beneath the cloak of folly, and the people gather them together; but adepts should apply themselves to the separation of the nuggets from the dross." M. le Capitaine d' Arpentigny has an almost parallel passage in his work, "La Science de la Main," and so it must be with us. In the following pages I have carefully excluded every indication of which I have not, by my own repeated experiences, recognized the reliability; it is only after experience that the Cheirosophic student will be able to decipher these apparently inexplicable signs. It is in this manner that Cheiromancy has survived all the other so-called "arts of divination"; many of the older Cheiromants turned their attention to other occult sciences,[79] but their works on Cheiromancy are the only ones which command attention at the present day.

¶ 92.
Indications of the body.

The time has come when the sciences which aim at reading the character of man from the indications of his body have become the objects of assiduous and scientific study, and of carefully directed research. At the end of the last century Johann Casper

[79] Thus Cocles and Tricasso both wrote works on Geomancy, or the practice of divination by the chance coincidences of objects, viz.:—TRICASSO DA CESARI, "Geomantia di Pietro d' Abano nuovamente tradottata per il Tricasso Mantuano" (Curtio Troiano di Navo) 1542; and BARTHELEMY COCLES, "La Geomantia nuovissimamente tradotta" (Venice, 1550).

Lavater published to the world the results of his investigations in Physiognomy,[80] showing how the propensities and inclinations of man are written in the features of his face, and F. J. Gall and his pupil Dr. Spurzheim, following in his footsteps, devoted their labours to the study of phrenology or craniology, with the result of presenting to succeeding generations of *savants* their well-known treatises on the subject they had studied to such good effect.[81] These two sciences of physiognomy and phrenology, having thus been established on a firm basis, it was only natural that the claims of the Hand should receive as much attention as those of the Face and Head,[82] in a word, that Cheirosophy should be lifted out of the slough of fanatical mysticism into which it had fallen, should be freed from the clouds of superstition and of charlatanry, which hid the nugget of pure gold amid the amalgam

Lavater.

Gall, Spurzheim.

Purification of the science.

---

[80] J. C. LAVATER, "Physiognomische Fragmente zur Beförderung der Menschenkenntniss und Menschenliebe" (Leipzig, 1775-8), 4 vols., 4to. *Translated into French:* "Essais sur la physionomie, destinés à faire connoître l'homme et à le faire aimer" (traduit en Français par Madame de la Fite et MM. Caillard et Henri Renfner) (La Haye, 1781-7 and 1803), 4 vols., 4to; *And into English:* "Essays on Physiognomy designed to promote the knowledge and the love of mankind" (translated from the French, by H. Hunter, D.D.), (London 1789-93), 3 vols., 8vo.

[81] F. J. GALL, "Anatomie et Physiologie du système nerveux en général, et du cerveau en particulier, avec des observations sur la possibilité de reconnaître les plusieurs dispositions intellectuelles et morales de l'homme et des animaux par la configuration de leurs têtes" (Paris, 1809-11-18-19), 4 vols., 4to, plates in fol : *and* " Sur les fonctions du cerveau et sur celle de chacune de ses parties, avec des observations sur la possibilité de reconnaître les instincts, les penchans, etc, . . . des hommes et des animaux par la configuration de leur cerveau et de leur tête " (Paris, 1822-5), 6 vols., 8vo.

[82] " La main est, aussi bien que les autres membres du corps un objet de la physionomie." " L'Art de Connaître les Hommes par la Physionomie" par GASPARD LAVATER. Nouvelle édition par M. Moreau. (Paris, 1806), 5 vols., 4to (vol. iii., p. 1).

of baser metal, and should claim its right to rank as a science worthy the consideration of high intellects, and the unbiassed judgment of men capable of expressing a sterling opinion. It is, therefore, the duty of the student of Cheirosophy to winnow the wheat from the chaff, and from the superabundant literature of the subject, from practical experience, and from personal observation, to deduce the scientific bases of the art of Cheirosophy; and to account for the fact that in the formations of, and in the signs visible in, the human hand an expert Cheiromant sees revealed the past life and present circumstances of the possessor of that hand, the tendencies of his nature and the future to which those tendencies, if unchecked, will lead him.

¶ 93
Arguments in favour of the study.

The arguments which may be adduced in favour of the study of the subject, are the same as may be brought forward in support of almost any science which may be stigmatised—not as occult—but as super-physical; and they may be found in most works which deal with psychic phenomena. They have never, perhaps, been more clearly stated by anyone

Roden Noel.

than by the Hon. Roden Noel,[83] who states fairly categorically most of the methods of reasoning which will lead to a just appreciation of these studies. He points out the leading fact that, though many people *feel*

Conviction.

the truth of what he calls spiritual laws, their brains suggest difficulties which they find it impossible to get over. *À propos* of the investigation of

Trickery.

trickery and the like agencies, which have always stood in the way of such study, he very pertinently remarks that one's own evidence ought not to be relied upon, unless one is accustomed to this species of investigation; but that we ought to rely upon the

Authorities.

experiences of such leading men as Crookes, Barrett,

[83] RODEN NOEL, " Philosophy of Immortality " (London, 1882).

Varley, and Zöllner, whose investigations, in which he was assisted by his colleagues Weber, Scheibner, and Fechner, afford a marvellous table of statistics for the student.[84] It is, however, of no real use to keep continually registering striking instances of the reliability of a science such as cheirosophy as shown by individual cases; the registration of such cases only proves the *fact* of their existence [which needs no support, their very existence affording their own proof], without explaining, in any way, the principles which lead to the existence of the fact; it is the principles which we ought, and which I have endeavoured in this argument to establish. Still, it is a leading axiom of research that new truths have ever to run the gauntlet between ridicule and persecution [*vide* ¶¶ 71 and 78ª], and discoverers are always regarded as foolish or fraudulent, until the absolute certainty of their discoveries has been established.

Registration of cases.

"Religion," says the scholiast, "has been given to men by revelation," and it has been the mental activity of the founders of religion which has caused it to take its acknowledged place; why, therefore, should we keep the principia of a new study [eclectic though it may be] bound hand and foot in the swaddling clothes of mental stagnation. Many there are who refuse to exert their brains on the behalf of eclectic science, who refuse to do the simplest things to test the truth of the dicta which they contemn. Are not such people like unto Naaman the Syrian, who "was wroth and went away" when he was commanded to do a simple thing, to demonstrate the power of the God of the prophet Elisha? The whole history of science is a record of great discoveries, arising from trivially simple circumstances; let us instance the kettle-lid, and the falling apple and the copper hook which led

¶ 94. *Revealed sciences.*

---

[84] A summary of the labours of Zöllner may be found in C. C. Massey's "Transcendental Physics" (London, 1880).

JamesWatt, Isaac Newton, and Galvani to the discovery of steam, the laws of gravitation, and of galvanism.[84a]

¶ 95. Birth of chemistry and astronomy.

We must remember that chemistry and astronomy had both their origin in pursuits which were looked upon as evil and occult sciences, and we may, I think, safely say with Honoré de Balzac, "In the present day so many established and authentic *facts* have been developed from occult sciences, that one day these sciences [*i.e.*, those of so-called divination] will be professed as nowadays men profess chemistry and astronomy!"[84a]

¶ 96. Ignorant incredulity.

"The self-complacent stolidity of *lazy* incredulity," says Noel, "is invincible," and this is epigrammatically true; the jeers of ignorant sceptics remind me always of the objectionable youth in the gallery at the playhouse, who, disapproving [he knows not why] of the performance, or not understanding the pabulum offered upon the stage to his debilitated intellect, finds expression for his feelings in profanity and "cat-calls." The laughter of people who have not given to this science a moment's rational thought, is to me as reasonable as if persons who have been born blind were to laugh at people who have that gift of sight of which they know nothing; the keen-sighted person only pities the blind man, and in like manner. . . . ? How much less cause still would the blind man have to laugh if his blindness resulted only from his own laziness which *refused* to see. Of course, a great factor in the ridicule and abuse with which the science meets is the incapacity which exists in a majority of minds to grasp the complex combinations of a science so deep as Cheirosophy. A perfect illustration of this is Herbert Spencer's old gentleman, who expresses authoritatively his

Incapacity to comprehend.

H. Spencer's illustration.

[84a] Some most interesting notes on these points may be found in Sir Lyon Playfair's "Speech on the second reading of Mr. Reed's Bill for the total suppression of scientific experiments upon animals" (London, 1883), subsequently printed in pamphlet form.

AN INTRODUCTORY ARGUMENT. 89

disapproval of classical music and preference for airs of the "Polly, put the kettle on" and "Johnny comes marching home" order, and whose mental state Herbert Spencer thus sums up:[85]—"On contemplating his mental state, you see that, along with absence of ability to grasp complex combinations, there goes no consciousness of the absence; there is no suspicion that such complex combinations exist, and that other persons have faculties for appreciating them."

"If one could *account* for the thing!" says the sceptic, and yet how many things he accepts that he cannot account for. Take, for instance, the every-day "phenomenon" of speech. He can examine by acoustics the vibrations of the air, he can examine by anatomy the tympanum of the ear and the otolithes, and by histology the auditory nerves, and the brain; but the minutest examination will not account for the sensation of sound which results in the production of *language*. Let him examine *all* his nerves, he will find them to all appearances identical with one another; yet how can he account for the absolutely different functions which the nerves perform, the differentiated actions of the sensory, the motor, the visual, the auscultatory, and the olfactory nerves, and of those which produce the sensations of taste? The phenomena cannot, by modern sciences, be accounted for, they *are*, they simply ARE. So with the indications of the hand. "Only a moiety of science," says Herbert Spencer, "is exact science; only phenomena of certain orders have had their relations expressed quantitatively as well as qualitatively. Of the remaining orders there are some produced by factors so numerous and so hard to measure, that to develop our knowledge of their relations into the quantitative form will be extremely

¶ 97. Unnecessary explanations.

Differentiation of nerves.

---

[85] HERBERT SPENCER, "Study of Sociology" (London, 1873), chap. vi., "Intellectual subjective difficulties."

difficult, if not impossible.[86] But these orders of phenomena are not, therefore, excluded from the conception of science. In geology, in biology, in psychology, most of the previsions are qualitative only; and where they are quantitative, their quantitativeness, never quite definite, is mostly very indefinite. Nevertheless, we unhesitatingly class these previsions as scientific." The above arguments point to what I think may some day be the line of reasoning which will lead to the physiological explanation of Cheirosophy; but, even if the indications cannot be accounted for, it does not alter the fact that they absolutely ARE.

¶ 98. Materialistic opposition.

The MATERIALISM of the age stands in our way; the materialism which seeks to "account for" everything, and which commences the process by doubting everything; the materialism which bids fair to lead us to doubt the existence of mind, of soul, of everything.

Dugald Stewart. "Of all the truths we *know*," says Dugald Stewart,[87] "the *existence* of the mind is the most certain. Even the system of Berkeley concerning the *non-existence of matter* is far more conceivable than that nothing but matter exists in the universe." And this same thought has been expressed by Francis Bacon, who commences his essay "Of Atheism" [1612] thus:—"I had rather believe all the fables of the 'Legend' and the 'Talmud' and the 'Alcoran,' than that this universal frame is without a mind." Doubt of the existence of mind [soul] is one of the least evils of this "fashionable materialism." I have had it hurled at me, in defence of materialist views, that electricity and steam are the outcomes of nineteenth-century materialism; but I have also heard it very ably debated that these very forces are ruining us by the facilities they afford for living at a headlong pace, that is

[86] HERBERT SPENCER, Op. cit., chap. ii., p. 45.
[87] DUGALD STEWART, "Elements of the Philosophy of the Human Mind" (London, 3 vols, 4to, 1792, 1814, and 1827).

rapidly whirling us towards a vortex of over-civilized annihilation.

¶ 99. "Honesty" of the science.

The dispute as to whether this science is "genuine" or not is simply a scandal. As I have said in another place,[88] "If this our science is a monstrous piece of assurance, of charlatanry, and of deceit, or if, at best, it is only a metaphysical amusement for a few minds, greedy of the marvellous and mystic, it cannot fail to be, in its very nature, repugnant to the soul of every thinker worthy the name, and of every student with the remotest inclination towards a love of the innately true. *But*, if, on the other hand, it possesses, as it must appear to do to any mind capable of an analysis of the study, the attributes I have claimed for it above, does it not deserve to rank as one of the highest and purest sciences which it has been vouchsafed to man to place himself in a position to comprehend?"

¶ 100. Its progress.

"It has been going on for so long, this old science," you say. Yes, it has indeed; but who shall say that it has not progressed? is not progressing? The world of thought laughs considerably less than it formerly did at this science, which it used to class with the other mystic or occult arts which cannot approach it in reasonableness, and which cannot produce for their support one single proof or hypothesis, where Cheirosophy can produce a thousand. The philosopher is not nearly so certain in his condemnations as he was formerly; the march of enlightenment and the love of progress have carried with them the noise of this Science of the Hand, till it has been so powerfully represented that the world is commencing to stand still, to let the science assert itself for what it is worth, and then to laud or damn it, as the case may be; is commencing to turn its attention towards the investigation

---

[88] "Codex Chiromantiæ," Codicillus, I. (London), Odd Volumes Opuscula, No. VII., 1883, p. 46.

of this science which is rising to the top of the crucible of knowledge, whence its dross and foreign impurities are passing away, volatilized by the blast-furnace of Common Sense.

¶ 101. Obstinate incredulity. There is also an astounding tendency among sceptics to declare, if they can find nothing else in the way of objection, that the "subject" is "in the swim," and agrees with the Cheirosophist so as to produce an effect upon a credulous audience.[89] Such ignorant incredulity as this is beneath our notice; but scientific and rational scepticism is the most valuable factor in the establishment of the science, for it shows us where our case requires strengthening, and by its reasonable counter-hypotheses shows us where we can strengthen the line of argument, which thus becomes every day stronger and stronger. *Fas est ab hoste doceri!*

¶ 102. Value of investigation. Again, it is a common thing for me to hear people say that this is a subject unworthy the attention of scientific men on account of its triviality. Unfortunately, the opinion expressed in Bacon's essay "Of Superstition," viz. :—"The master of superstition is the people, and in all superstition wise men *follow* fools, and arguments are fitted to practise in a reverse order," was a painfully correct one, and its truth is evident even to-day. Wise men will not, as a rule, take the trouble to investigate even so pertinent a matter as

---

* When Dr. Esdaile made his now celebrated experiments and investigations in mesmerism as an anæsthetic in the Calcutta hospital, cutting out tumours, amputating limbs, and performing terrible operations on patients who had merely been thrown into an hypnotic trance, without the administration of any other anæsthetic, there were not wanting persons who stated that the patients *had been bribed not to scream under the agonies that they suffered.* A full account of these most fascinatingly interesting experiences may be found in Dr. Esdaile's works, "Mesmerism in India, and its practical application in Surgery and Medicine" (London, 1846), and "The Introduction of Mesmerism (with the sanction of Government) into the Hospitals of India" London, 2nd edition, 1856).

Cheirosophy, and what can be more mistaken than this neglect? Once allow that the claims of the science are genuine, and we have before us a science of the most astounding importance to humanity; let it, therefore, be investigated, and if investigation proves that the science is chimærical, the time which will have been thus employed will *not* have been wasted, but will have been most profitably spent in the timely suppression of a vast and degrading illusion.

¶ 103. Narrow-minded arguments of religion and science.

We know, all of us, by bitter experience, that religion, and even on occasion science, will denounce and deny not only theories, but even *facts*, when they are opposed to them. [Thus, for instance, at the end of the last century, it was firmly *denied* that aërolites fell, or could fall, from the sky, and the idea was utterly ridiculed.] The term "scientific impossibility" is too often nothing more or less than a synonym for "reasonable improbability." A writer in the *Allgemeine Zeitung*, in March 1884, summed up these points with much conciseness, pointing out the counter-danger that once people begin to be convinced they are painfully apt to take everything and anything upon trust; he sums up his discourse by saying:—"In our opinion, *superstition* will only come to an end when exact science will take the trouble to examine, without prejudice, the facts it has hitherto distinctly denied; that is to say, when it will approach them with the admission that things are not necessarily untrue, because they are unexplained."

¶ 104. Obstacles to the study.

The causes of psychological ignorance are principally these: the imperfections of language, which prevent people from explaining and defining what they really mean; the tendency to grasp general principles without studying particular facts; the extreme difficulty we experience in getting at any accurate description of mental conditions; the prejudices which arise from reverence for great opposing authorities and local traditions; and a deeply rooted *penchant* to run after, and, so to speak, to "swallow whole" any opinions, etc., which may be in their very nature singular or paradoxical, a readiness to believe improbabilities, like that of the Levantine family, who insisted on believing that "The Arabian Nights" were true histories.[90]

¶ 105. Conclusion. Now I have done. *Experto crede!* "I have said all that I think necessary, and trust you will adopt that course which is best for the community and for yourselves."[91] Believe, I pray you, intelligent reader, that the science I have done my best to place before you, in its highest state of development, in the following pages, and which I have endeavoured to explain and prepare you for in the foregoing argument, is no idle pastime, is no frivolous *passe-temps*, is no exercise of assurance; but that this little work is a simple manual towards the interpretation of one of the pages of the greater Book of Nature; and that, seeing yourself as no one else sees you, and understanding the characters of those among whom your lot in life is cast, you may, in promoting the welfare of the individual, be advancing the welfare of the community; and that you may appreciate the privileges, with a due regard for the responsibilities of the Cheirosophist.

July 13th, 1885.

[90] BAYLE ST. JOHN, "Two Years' Residence in a Levantine Family" (London, 1850).
[91] ΔΕΜΟΣΘΕΝΟΤΣ ΟΛΤΝΘΙΑΚΟΣΓ' (36). Σκεδὸν εἴρηκα ἃ νομίζω συμφέρειν, ὑμεῖς δ᾽ ἕλοισθε ὅ τι καὶ τῇ πόλει καὶ ἅπασι εὐνοίσειν ὑμῖν μέλλει.

SECTION I.

Cheirognomy.

1. J. B[ULWER]—"Chirologia or the Naturall Language of the Hand, composed of the Speaking Motions and Discoursing Gestures thereof, whereunto is added Chironomia, or the Art of Manuall Rhetoricke," etc. (London, 1644.)
2. S. D'ARPENTIGNY—"La Chirognomonie; ou l'Art de reconnaître les Tendences de l'Intelligence, d'après les Formes de la Main." (Paris, 1843.)
    "La Science de la Main; ou l'Art de reconnaître," etc., etc. (Paris, 1865.) *3rd edition.*
3. ANONYMOUS — "The Hand: phrenologically considered, being a glimpse at the relation of the Mind with the Organization of the Body." (London, 1848.)
4. R. BEAMISH—"The Psychonomy of the Hand, or the Hand an Index of the Mental Development." (London, 1865.)
5. C. WARREN—"The Life-size Outlines of the Hands of Twenty-two Celebrated Persons." (London, 1882.)
6. E. HERON-ALLEN — "Codex Chiromantiæ, being a compleate Manualle of ye Science and Arte of Expoundynge Ye Past, Ye Presente, and Ye Future and Ye Charactere by Ye Scrutinie of Ye Hande, Ye Gestures thereof and Ye Chirographie."—Codicillus I. Chirognomy. (London, 1883.) *Odd Volumes Opuscula, No. vii.*
7. R. BAUGHAN—" Chirognomancy or Indications of Temperament and Aptitudes manifested by the Form and Texture of the Thumb and Fingers." (London, 1884.)

SECTION I.

CHEIROGNOMY; OR, THE SHAPES OF THE HANDS.

'Η μὲν οὖν ὑπόσχεσις οὕτω μεγάλη, τὸ δὲ πραγμ' ἤδη τὸν ἔλεγχον δώσει, κριταὶ δ' ὑμεῖς ἔσεσθε.
ΔΗΜΟΣΘΕΝΟΥΣ ΚΑΤΑ ΦΙΛΙΠΠΟΥ Α'.

FOREWORD.—It is usual to divide the science of Cheirosophy into two principal sections: Cheirognomy, or the science of interpreting the characters and instincts of men from the outward formations and aspects of their hands; and Cheiromancy, or the science of reading the characters and instincts of men, their actions and habits, and the events of their past, present, and future lives, in the lines and formations of the *palms* of their hands. Though, as will be seen anon, the line of demarcation which has been drawn between these two branches of the science is not only false in principle but misleading in practice, [for, as will be seen in the following pages, the two sections are inextricably intermingled, and cannot be separated if accuracy of result is aimed at,] it is still convenient to preserve the semblance of separation, so that the student may master the principia of cheirognomy before he begins to apply it to the interpretation and elucidation of the more intricate rules of cheiromancy, and for this reason

¶ 106.
The two branches of Cheirosophy.

I have divided the one great subject of cheirosophy into its two constituent and companion elements of cheirognomy and cheiromancy.

¶ 107. Cheirognomy. Cheirognomy, therefore, is that branch of the science of the hand which enables us, by a mere superficial observation of the exterior formations and appearance of the hands, and by the impressions produced by them on the senses of vision and of touch, to arrive at an accurate estimate of the character, disposition, and natural propensities of any individual in whose presence we may find ourselves. It is of the highest importance that the student of cheirosophy should first master this very important branch; for what is more obvious than that the character and tendency of the mind and the natural inclinations of the subject[92] under examination should so materially influence his actions, manner, and speech, his physical and moral bases of life, and the events of his existence, that by getting at the former by the aspect of his hands the knowledge of the latter follows almost of itself?

¶ 108. Cheirognomy and Cheiromancy. Again, it will be borne in mind that the cheirognomy of a subject—that is to say, the shape of his (or her) hands—is often hereditary and inborn, the physiological legacy of a long line of ancestors, whose characters and peculiarities of mind he may possibly inherit (as Aristotle has pointed out in his treatise on the "Generation of Animals"[93]), whilst the lines, signs, and mounts of the palm—that is to say, the cheiromancy of a subject—are more often the results of the external and internal influences, such as the astral and

---

[92] By the word "*subject*" throughout this work, I mean the person whose hands are, for the moment, under examination.

[93] ARISTOTLE: ΠΕΡΙ ΖΩΩΝ ΓΕΝΕΣΕΩΣ. βιβλ. Λ'; Κεφ. ιή. Προσήκει δὲ μᾶλλον ἀπ' ἐκείνων· πρότερα γὰρ ἐκεῖνα καὶ σύγκεται τὰ ἀναμοιομερῆ ἐξ ἐκείνων καὶ ὥσπερ πρόσωπον καὶ χείρας γίγνονται ἐοικότες, οὕτω καὶ σάρκος καὶ ὄνυχας.

cerebro-nervous fluids to which I have adverted at length in the Introductory Argument.

Cheirognomically speaking, hands are divided into seven classes or types, each of which will in due course receive careful attention; firstly, however, it is necessary to consider the interpretation of the many general features of a hand, which carry with them their own significations, to whatever type that hand may belong.

¶ 109. The divisions of Cheirognomy.

SUB-SECTION I.

CONCERNING THE HAND IN GENERAL AND THE INDICA-
TIONS AFFORDED BY THE ASPECTS AND CONDITIONS
OF ITS VARIOUS PARTS IN PARTICULAR.

¶ 110.
The Seven
Types.
To whatever type a hand may belong, there are certain aspects and formations of its constituent parts which materially affect the tendencies indicated by the development of that particular type, and these aspects and conditions must be carefully considered in the preliminary examination of that hand. Such are the developments and formations of the palm, the fingers, the joints, the thumb, the relative size and proportions of the whole hand and of its constituent parts, all of which matters must be observed carefully, to arrive at the true influence of the developed, or mainly developed, type; and to explain the indications which are read in these circumstances and conditions is the aim of the present sub-section.

THE PALM.   § 1. *The Palm of the Hand.*

¶ 111.
Its Indications.
In the first place, you will notice the formation and the physiological conditions of the *palm*. In it are

found the *physical* attributes of the character and the intensity with which they are developed.

If the palm is thin, skinny, and narrow, it indicates timidity, a feeble mind, narrowness and paucity of intellect, and a want of depth of character, energy, and moral force.[94]

¶ 112. Thin and narrow.

If, on the other hand, it is in perfect proportion with the fingers, the thumb, and the rest of the body, firm without being hard, elastic without approaching to flabbiness, the mind thereby indicated is evenly balanced, ready to accept impressions, appreciative, intelligent, and capable of sustaining and directing the promptings of the instinct. If, however, this last hand is too highly developed, and its proportions are too strongly accentuated, the exaggeration of these qualities tends to produce over-confidence, selfishness, and sensuality; whilst if, going a step farther, the hand joins to these highly-developed proportions a hardness and resistance to the touch, and the palm is longer than the fingers, the character tends towards brutality of instinct, and a low grade of intelligence is betrayed by the animality of the ideas. [It will be noted *infrâ* [¶ 265] that these last characteristics are those *par excellence* of the elementary type.]

¶ 113. Well-proportioned

Over-developed

Hardness.

A hollow, deep palm denotes almost invariably misfortune, loss of money, misery, and danger of failure in enterprise. [It will be seen *infrâ* that this is caused by a defection of the Plain of Mars (*vide* ¶ 380, and p. 217), and is a sign of ill luck even when the rest of the hand is favourable.]

¶ 114. Hollow palm.

The palm, therefore, must be absolutely normal, and naturally proportioned to the rest of the hand (*i.e.*, to the thumb and fingers), and thus to the rest

¶ 115. Necessity of normal condition

[94] This is also one of the rules laid down by Aristotle in his treatise on Physiognomy : ΦΤΣΙΟΓΝΟΜΙΚΑ, Κεφ. γ´ : Δειλοῦ σημεῖα . . . χεῖρες λεπταὶ καὶ μακραί.

of the body.⁹⁵ In any other case its indications will be found to modify those of the rest of the hand, to the consideration of which we can now turn.

¶ 116.
Excesses of formation.

Any excess in the formation of any part of the hand is bad, denoting disorder and demoralization of the qualities indicated by the formation which is in excess, and this is the more infallible [as will be seen] if the phalanx of the thumb, wherein are seated the indications of *the will*, be long.

THE FINGER JOINTS.

§ 2. *The Joints of the Fingers.*

¶ 117.
Smooth and jointed fingers.

Looking at the fingers of the whole world, they divide themselves, cheirognomically speaking, into two great classes: (α) Fingers which are knotted, and (β) fingers which are smooth; that is to say, (α) those in which the joints are so developed as to cause a perceptible "bulge" where they occur between the phalanges of the fingers, and (β) those in which the joints are so little pronounced as to be imperceptible at first sight; and the former class divides itself again into two sub-classes: (α 1) those fingers which have both joints developed, and (α 2) those which have but one.

¶ 118.
The joints.

Development of the joints of the fingers indicates thought and order, which are greater or less in their influence on the life, according as one or both joints are to a greater or less degree prominent.

¶ 119.
The upper joint.

If the first joint [*i.e.*, that which connects the first (or nailed) phalanx and the second (or middle) phalanx] is developed, accentuating the junction of the first and second phalanges of the fingers, it indicates a method and reason in the ideas, a well-ordered mind, and a neat administrative disposition. The development of this joint, if the phalanx of will [on the thumb] is long, is generally indicative of remarkable intelligence;

⁹⁵ Palma truncata et digitis impar indicat in manu fœminæ difficilem et periculosam lucinam.

THE JOINTS OF THE FINGERS.

but if the phalanx of will is short, this development of the first joint often betrays excess of ill-directed reasoning, tending to paradoxicalism, and this is more certainly the case [as will be seen *infrâ*, ¶ 583] if the Line of the Head decline upon the Mount of the Moon and the fingers are pointed.[96] When the Mount of Jupiter is high in the hand, the development of this joint denotes vanity.

¶ 120. The upper joint unsupported

If this first joint be very prominent there is always a great deal of talent in the subject; but if the lines of the palm are thin and dry, and the thumb is small, a lamentable want of soul is generally apparent. Reason, however, remains always the prevailing instinct.

¶ 121. Both joints developed.

If the second joint [*i.e.*, that which connects the second (or middle) phalanx and the third (or lower) phalanx] is also developed, the instincts of reason and order are the more strongly pronounced. In this case the prevailing instincts of the subject will be symmetry, order, and punctuality. The mind will be well regulated, the ideas will be good and equitable, and the actions will be governed by reflection and deliberation. There will be the love of analysis and of inquiry, and a strong *penchant* towards the sciences. Both joints thus developed, and the Mount of the Moon high in the palm, indicate a love of poetry and of music, but the poetry must be grand and reasonable (not fantastic or erotic), and the music will be scientific and true [harmony, counterpoint, fugue, and the like].

Influence of the moon.

[96] The reader is requested not to take alarm at sentences like the foregoing, which at present must seem to him to be unintelligible, and savouring of astrology and charlatanry. The reason of their use, as well as their meaning, will become quite clear to him as he reads the following pages, and as I have said *suprâ*, ¶ 106, it is useless and impossible to endeavour to separate wholly the two branches of cheirosophy known for distinction as cheirognomy and cheiromancy.

¶ 122.
The ower joint. The development of the second joint only, gives to a subject order and arrangement in things material and worldly, as opposed to the orderliness in things mental and psychological, which is indicated by the development of the first [or upper] joint. The orderliness of the second joint is that which appertains to things connected with one's *self*, a selfish order which produces merchants, calculators, speculators, and egoists.

¶ 123.
Smooth fingers. If on the other hand your fingers have *neither* joint highly developed, [*i.e.*, no perceptible bulge is to be seen at the joints,] your *penchant* will be towards the arts. Your proceedings and actions will be governed by inspiration and by impulse, by sentiment and by fancy, rather than, as in the former case, by reasoning, knowledge, and analysis, and whatever the type of the hand, if the fingers are smooth, the first impression of that subject is always the correct one, and subsequent reflection will not help him in arriving at a conclusion.

¶ 124.
Smooth fingers with upper joint perceptible. Smooth fingers with the first joint indicated by a bulge which is not very much accentuated, often denote a talent for spontaneous invention and intuition in the pursuit of science, but these qualities are never in this case the result of calculation. This first joint *rising* only on the back of the fingers, *not* bulging out at their sides, indicates a talent for invention.

¶ 125.
Bad line of head. When with smooth fingers the Line of the Head [*vide* ¶¶ 572 and 583] is bad and twisted, declining upon the Mount of the Moon, which is high, with a short phalanx of logic in the thumb, though the intuition remains, it will generally be all wrong, and give to the subject the most false conceptions.

¶ 126.
Effect of joints. Thus it is easily explained that whilst knotty-fingered [*i.e.*, prominent-jointed] subjects have most *taste* intellectually speaking, [taste, properly so called,

THE JOINTS OF THE FINGERS.

being born of reason and intellectual consideration], those with smooth fingers have the larger share of natural and unreasoning grace. Passion [as opposed to sensuality] is the *worldly* instinct of the former, whereas sensuality [as opposed to passion] is generally a characteristic of the latter. Effect of smooth fingers.

By a like chain of argument, smooth-fingered subjects often fail in their undertakings through pursuing them too hotly and impulsively, and when with smooth fingers the Line of Head is separated from the Line of Life [*vide* ¶ 582] the badness of the latter sign is the more pronounced, for the impulse of the smooth fingers will carry into prompt and unconsidered action the false impressions and mistaken self-confidence of the separated lines. ¶ 127. Smooth fingers.
Bad line of head.

Throughout the examination of hands, these two principia must be borne in mind—that the jointed subject works by calculation, reason, and knowledge, whilst the action of the smooth-fingered subject is born of, and governed by spontaneity, instinct, impulse, and inspiration. ¶ 128. Differences of impulse.

At the same time one must never lose sight of this particular; that, though with the first joint developed a hand may betray artistic instincts, if both joints are prominent, art becomes a thing tolerated merely, and not a thing understood. ¶ 129. Effect of joints.

Education, self-discipline, and cultivation may develop joints in a hand, and may cause fingers originally rounded to become square, or even spatulated, but they can never erase the joints and produce a smooth-fingered hand, or mould square or spatulated fingers into roundness, for it is easier to go from artistic to scientific instincts, from intuition to knowledge, or from idealism to materialism, than *vice versâ*. ¶ 130. Development and disappearance of joints.

§ 3. *The Comparative Length of the Fingers.*

LENGTH OF THE FINGERS.

Again, the fingers of a hand are either short or long. That is, on first sight they may strike one as being either short or long by comparison with the palm and rest of the hand, or by comparison with the majority of fingers one is in the habit of seeing.

¶ 131. Short fingers.

People with short fingers are quicker, more impulsive, and act more by intuition and on the spur of the moment, than people with long; they prefer generalities to details, jumping hastily to conclusions, and are quick at grasping the entirety of a subject.

¶ 132. Effect of short fingers.

They are not particular about trifles, caring little for appearances and for the conventionalities of life; but their leading feature is their quickness of instinct and action. Their judgment is quick, and their action is prompt, and they have, to a remarkable degree, the instinct of the perception of masses. They are brief and concise in expression and in writing, but often when the rest of the hand is weak such subjects are given to frivolity and chattering.

¶ 133. Thick and short fingers.

Weak hand.

If the fingers are thick as well as short it is a sign of cruelty. Short fingers with a short line of head denote want of tact, and carelessness in acting on impulse, especially if the Mount of the Moon is highly developed; but with short nails and a long line of head, the instinct of synthesis [which is the great attribute of the short-fingered subject] gives a talent for grasping particulars and comprehending a scheme which produces a rare faculty for administration.

¶ 134. Short fingers with joints.

If with short fingers either or both of the joints are developed, they will have a certain amount of reason and calculation to assist the quickness of their intellect, which will thus be supplemented by a powerful auxiliary, for the calculation indicated by the joints will be able to apply itself with the rapidity of comprehension indicated by the shortness of the fingers.

With long fingers we find a love of detail even to frivolousness, an instinct of minutiæ which often blinds the subject to the appreciation of the harmonious whole, carefulness in dress and behaviour, and consequent hate of slovenliness or brusquerie of manner. Such a subject will be respectful and dignified, easily put out, and easily pleased by an attention to the minor peculiarities of his nature. ¶ 135. Long fingers.

If long fingers have the first joint developed, such a subject will be inquisitive, watchful, always on his guard against liberties, observant of small things, and addicted to manias and idiosyncracies about things, especially if the phalanx of logic in his thumb be long. ¶ 136. Long fingers and upper joint.

Artists with such fingers as these will often elaborate detail at the expense of the mass of the subject upon which they are working, and all persons whose fingers present this formation will be distrustful, always trying to seek out second meanings for one's remarks, and attributing motives and deep significations to one's most meaningless speeches and most trivial actions. ¶ 137. Long fingered artists and others.

Long fingers, therefore, betray a worrying disposition, worrying both to themselves and to others, unless a long Line of Head and a well-developed phalanx of will modify the indications of the fingers. ¶ 138. Effect of long fingers.

In literature such subjects pay an attention to detail which is maddening to see in print; for they go off at a tangent, and discourse on matters more or less germane to the subject in hand, until one loses sight of the prime object of the argument, which thus becomes confused and wearisome. ¶ 139. Long fingers in literature.

Such hands, also, often betray cowardice, deceitfulness, and affectation; but these tendencies may be overruled by a good Line of Head and a well-developed Mount of Mars. [*Vide* ¶¶ 476-7.] ¶ 140. Bad effects of long fingers.

¶ 141.
Long fingers and both joints.
With both joints developed you will find pugnacity, argument, and a didactic mode of expression, boldness of manner and speech, and even malice, especially when to these long jointed fingers a subject adds a large thumb, which indications generally reveal chicanery, dishonesty, a controversial humour, and a penchant towards scandal and mischief making; the latter particularly when the fingers terminate in short nails.

¶ 142.
Large, medium, and small hands.
Thus, to recapitulate: a large hand indicates a love and appreciation of details and minutiæ; a medium-sized hand denotes comprehension of details *and* power of grasping a whole; whilst very small hands betray always the instincts and appreciation of synthesis.

¶ 143.
Differences between large and small-handed subjects.
The large-handed subject will have things small in themselves, but exquisitely finished, whilst the small-handed subject desires the massive, the grandiose, and the colossal. Artists in horology have always large, whilst the designers and builders of pyramids and colossal temples have always small hands. In Egyptian papyri and hieroglyphic inscriptions the smallness of the hands of the persons represented always strikes one at first sight.

¶ 144.
Handwritings.
In like manner people with small hands always write large, whilst people with large hands always write [naturally] small.

¶ 145.
Medium hands.
Thus it will be seen that it is only medium-proportioned hands that possess the talents of synthesis *and* of analysis, the power of appreciating at the same time the mass, and the details of which it is constituted.

THE FINGERS GENERALLY.

§ 4. *The Fingers generally.*

¶ 146.
The three phalanges.
The three phalanges of the fingers have also their significations. Thus, the first phalanges of the fingers represent the intuitive faculties, the second phalanges represent the reasoning powers, and the third or

THE FINGERS GENERALLY. 109

lowest phalanges represent the material instincts. Thus, therefore, if the third phalanges are relatively the largest, and are thick and full by comparison with the others, the prevailing instincts will be those of sensuality and of luxury; if the second phalanges are the most considerable, a love of reason and reasoning will be the mainspring of the life, whilst with a high development of the first [or exterior] phalanges the intuitive and divine attributes will be the prevailing characteristics of the subject.

¶ 147. Effect of the joints.

Thus, it will be seen, the joints seem to form, as it were, walls between the worlds; the joint of philosophy and of reason dividing the phalanx of intuition and instinct from the phalanx of reason and knowledge; and the joint of material order forming the boundary betwixt the reasoning faculties and the world of materialism.

¶ 148. Thick fingers.

From what has gone before it will be comprehended that thick fingers will always denote a love of ease and luxury; but also unless the hand is hard the subject will not seek and require luxury; he will only enjoy and appreciate it when it comes in his way.

¶ 149. Twisted fingers.

When the fingers are twisted and malformed, with short nails, and only the elementary lines [those of head, heart, and life] are visible in the hand, it is almost infallibly the sign of a cruel and tyrannical disposition, if not of a murderous instinct; but if these twisted fingers are found on an otherwise good hand the deduction to be made will only be that of a mocking and annoying disposition.

¶ 150. Stiff and hard hands.

If a hand is stiff and hard, opening with difficulty to its full extent, it betrays stubbornness of character.

¶ 151. Fingers turning back.

People whose fingers have a tendency to turn back, being supple and elastic, are generally sagacious and clever, though inclined to extravagance, and always curious and inquisitive.

¶ 152.
Fingers fitting into one another or not.

The fingers fitting closely together without interstices between them denotes avarice, whereas if there are considerable interstices and chinks between them which show the light through when the hand is held between the eye and the light, it is a sign [like the turning back of the fingers] of inquisitiveness.

¶ 153.
Transparent fingers.

Smoothness *and transparency* of the fingers betray indiscretion and loquacity.

¶ 154.
Ball at the finger tips.

Whatever may be the formation of the fingers, the type to which they belong, or the other conditions of the hand, if a little fleshy ball or knob be found on the face of the first phalanx it is a sign of extreme sensitiveness and sensibility, of tact, [from the dread of inflicting pain upon others,] and of taste, [which is the natural heritage of a nature so gifted].

It may be noted, also, in this place that there are certain indications to be read in the greater or less length and development of each separate finger; but this I shall notice further on under the heading of the Cheirognomy of the Individual Fingers. [*Vide* p. 125.]

Gaule the sceptic.

His indications.

Gaule,[97] who never loses an opportunity of railing against Cheiromancy as it was practised in his day, in mentioning the science as it existed then, says, quoting the Cheiromants in derision :—" A great thick hand signes one not only strong but stout; a little slender hand one not only weak but timorous; a long hand and long fingers betoken a man not only apt for mechanical artifice but liberally ingenious; but those short on the contrary note a foole and fit for nothing; an harde, brawny hand signes dull and rude; a soft hand witty but effeminate; an hairy hande luxurious; long joints signe generous, yet if they be thick withal, not so ingenious. Short and fat fingers mark a man out for intemperate and silly; but long and leane for

---

[97] JOHN GAULE, "Πυσμαντία: the Mag-astromancer, or the Magicall-Astrologicall-Diviner posed and puzzled" (London, 1652, p. 187).

witty; if his fingers crook upward that shows him liberal, if downward niggardly"; and it will be interesting for the reader to note, during the perusal of the following pages, that these indications are quite correct.

Their correctness.

§ 5. *The Finger Tips.*

The Finger Tips.

The first [or exterior] phalanges of the fingers of a hand present four principal formations. They are either (a) "Spatulate," *i.e.*, the tip of the finger is broad and flat, or club-shaped, like the "spatula" with which a chymist mixes his drugs. (β) "Square," *i.e.*, the tip of the finger, instead of being round and cylindrical and curved over the top, is flat upon the tip, and so shaped that a transverse section of the tip would present the appearance of a square, at least as regards three sides thereof: [the inside of the finger tip is in almost all cases curved]. (γ) "Conic," *i.e.*, the tip is cylindrical, and rounded over the top like a thimble; or (δ) "Pointed," *i.e.*, the finger ends in a more or less extended circular point: and each of these forms has such marked and different characteristics as almost to constitute types by themselves. [With certain concomitant signs they do constitute the Types of Cheirognomy which will be fully considered in a future sub-section; but it seems right here to notice the particular instincts indicated by each one in particular.]

¶ 155. *The four principal formations.*

If your fingers terminate in spatule your first desire will be for action, activity, movement, locomotion, and manual exercise; you will have a love of what is useful, physical, and reasonable; yours will be the appreciation of things from the utilitarian point of view, love of animals, and inclination for travel, war, agriculture, and commerce. You will interest yourself principally in the things of real life—physical and mechanical force, calculation, industry, applied sciences, decorative art, and so on.

¶ 156. *Spatulate fingers.*

¶ 157.
Jointed or smooth spatulate fingers.

And here [to recede a little] you must take into consideration what we said about the joints (p. 102), understanding that the subject with spatulate *knotty* fingers will develop and pursue the propensities of the spatulate finger-tip by reason, calculation, and knowledge, as opposed to the subject with spatulate *smooth* fingers, who will develop the same characteristics by spontaneity, by impulse, by rapid locomotion, and by inspiration. Thus, if your fingers terminating in spatule have the joints developed you will excel in *practical* science and *scientific* mechanics, [such as statics, dynamics, navigation, architecture, and the like]. And [as we shall see presently] the tendencies of this spatulate formation of the finger tips are the more accentuated if you add to them a large thumb and firm hands.

¶ 158.
Square fingers.

If your finger tips are square your prevailing characteristics will be symmetry and exactitude of thought and habit. You will have a taste for philosophy, politics, social science and morals, languages, logic, geometry, [though you will probably only study them superficially]. You will admire dramatic, analytic, and didactic poetry, and you will require and appreciate metre, rhythm, construction, grammar, and arrangement in literature, whether poetic or otherwise, and your admiration in art will be for the defined and conventional. You will have business capacity and respect for authority, combined with moderate but positive ideas. You will incline to discovery rather than to imagination, to theory and rhetoric rather than to practical action.

¶ 159.
Tidiness.

You will admire order and tidiness, but unless your fingers have the joints developed, you will not practise the tidiness you admire—*i.e.*, you will arrange things that are visible, but your drawers and cupboards will be in confusion.

THE FINGER TIPS.

Of course, as before, the distinctions of the knotty and the smooth fingers apply to this formation of the finger tips [vide ¶ 157]; the former being always the more sincere and the more trustworthy—the more ready to put their theories into practice. As we shall presently see [vide ¶ 309], a high development of the joints, combined with a large thumb, will give to these square-tipped fingers the most fanatical red-tapeism, regularity, and self-discipline.

¶ 160. Smooth and jointed square fingers.

Thus it will be easily comprehended that between the spatulate and the square finger tips there are great distinctions, the principal being those of simplicity as opposed to politeness, and of freedom as contrasted with elegance.

¶ 161. Spatulate and square finger tips.

Amongst musical people the most thorough theoretical musicians have square fingers, by reason of the amount of rhythm and symmetrical exactitude required. Brilliant execution and talent as an instrumentalist is always accompanied by spatulate fingers, [which are *not*, as so many people imagine, the *result* of instrumental practice, but of the temperament which makes that practice a pleasure,] whilst singers [who are essentially melodists] have nearly always conical and sometimes pointed fingers.

¶ 162. Musical fingers

Again, if your fingers terminate conically, your whole instinct will be artistic. You will love art in all its branches, and adore the beautiful in the actual and visible form; you will be enthusiastic, and inclined to romance and social independence, objecting to stern analysis; your greatest danger is that of being carried away into fantasy.

¶ 163. Conical fingers.

If these fingers have either or both joints developed, you will have more moral force, and will be able to keep your more unruly instincts in control. And, as we shall see presently, the tendencies of this conical formation of the finger tips are the more accentuated if the subject have also soft hands and a small thumb.

¶ 164. Conic jointed fingers.

This remark also applies to the pointed formation of the finger tip next below noticed.

¶ 165.
Effect of the thumb.
When fingers of these formations [the conic and the pointed] are gifted with a large thumb, their instinctive art will expand itself logically and methodically, almost as if the finger tips were square.

¶ 166.
Pointed fingers:
And, lastly, suppose your fingers take the form of a cone, drawn out even to pointedness, yours will be exclusively the domain of ideality, contemplation, religious fervour, indifference to worldly interests, poetry of heart and soul, and yearning for love and liberty, cultivation [even to adoration] of the beautiful in the æsthetic abstract rather than in the visible and solid.

¶ 167.
Effect of pointedness.
Whatever may be the type, formation, or conditions of a hand, a pointed formation of the finger-tips will denote impressionability of the subject. This formation [like the others] will be considered at greater length under the heading of the type to which it particularly belongs.

¶ 168.
The four formations.
These are the four principal formations of the finger tips, concerning which space renders it impossible, and the intelligence of the average reader renders it unnecessary, to go further at this present.

¶ 169.
Amorphous finger-tips.
If the fingers cannot be classed under any of these formations, but have their tips absolutely shapeless, and consequently irresponsibly ugly and malformed, such a hand is that of a person whose intellect is weak, and whose individuality is practically *nil*.

¶ 170.
Excess of formations.
It must be borne in mind that *exaggeration* or *excess* of any form denotes a diseased condition of the instincts indicated, by reason of their too high development.

¶ 171.
Excessive pointedness.
Thus, an exaggerated pointedness is apt to be the result of impossible and fanatical romanticism, foolhardiness, and imprudence, exaggeration of imagina-

tion, which developes into lying, and particularly into affectation and eccentricity of manner.

Fingers too square show fanatical love of order and method in the abstract, servile submission to conventionality, and to self-prescribed and otherwise regulated ordinances. ¶ 172. Excess of squareness.

Exaggerated spatulation of the fingers indicates tyranny, [especially in the thumb, *vide infrâ*, ¶ 194,] perpetual hurry, restlessness, and discontent with one's fellow-creatures. ¶ 173. Excessive spatule.

These excesses of formation are also much influenced by the development or want of development of the thumb [*vide* ¶¶ 197-198 and 202-3]. ¶ 174. Effect of the thumb upon excesses.

§ 6. *The Hairiness of the Hand.* HAIRINESS.

To leave nothing connected with the hand unconsidered, the greater or less amount of hair found thereon must also engage our attention.

A hand the back of which is very hairy betokens inconstancy, whilst a quite hairless and smooth hand denotes folly and presumption. A slight hairiness gives prudence and love of luxury to a man; but a hairy hand on a woman always denotes cruelty. ¶ 175. Hairy hands and smooth hands.

Hair upon the thumb [according to the Sieur de Peruchio] denotes ingenuity: on the third or lower phalanges of the fingers only, it betrays affectation, and on all the phalanges a quick temper and choleric disposition. ¶ 176. Hairy thumbs and fingers.

Complete absence of hair upon the hands betokens effeminacy and cowardice. ¶ 177. Absence of hair.

§ 7. *The Colour of the Hands.* COLOUR.

If the hands are continually white, never changing colour [or only doing so very slightly] under the influences of heat or of cold, they denote egoism, selfishness, and a want of sympathy with the joys and sorrows of others. ¶ 178. Constantly white hands.

¶ 179.
Persons whose colorations are significant

Le Sieur de Peruchio observes, very truly,[98] that in cases such as those of soldiers, of servants, and of workpeople, whose daily occupations must necessarily alter and affect the coloration of their hands, the colours cannot be relied upon as a certain indication of the temperament; but in the case of women and of persons whose sedentary habits, whose light occupations, or whose care of their hands, tend to preserve them in their normal and natural colours and conditions, the following data may with confidence be gone upon.

¶ 180.
Red, yellow, and dark hands.

Redness of the skin denotes sanguinity and hopefulness of temperament; yellowness denotes biliousness of disposition;[99] blackness, melancholy; and pallor, a phlegmatic spirit.

¶ 181.
Preferable tint.

Darkness of tint is always preferable to paleness, which betrays effeminacy; the best colour being a decided and wholesome rosiness, which betokens a bright and just disposition.

THE THUMB.

§ 8. *The Thumb.*

¶ 182.
Importance of the thumb.

I have in the Introductory Argument [*vide* ¶¶ 26, 31, 35] said so much concerning the importance of the thumb and its position in the physiological œconomy of the hand, that it is needless to repeat any of those things in this place. It is sufficient to say that the thumb

---

[98] "La Chiromancie, La Physionomie et la Géomancie, avec la signification des nombres, et l'usage de la Roue de Pythagore," par LE SIEUR DE PERUCHIO. (Paris) *P. L'Amy.* 1657. 4to.

[99] BARTEN HOLYDAY, in his "Τεχνογαμία; or, the Marriage of the Arts" [a Comedie: London, 1618], says, by one of his characters, "That a *yellow death mould* may never appeare upon your Hande." The colour yellow seems always to have been regarded as ominous [*vide* ¶ 262]. We find in an old play in Dodsley's Collection [Edition of 1780, vol. vi., p. 357] the following passage:—

"When yellow spots do on your Hands appear,
Be certain then you of a corse shall hear" . . .

is by far the most important part of the hand, both cheirognomically and practically speaking, for without it the hand would be comparatively [if not absolutely] powerless, and in it the cheirosophist looks for the indications of the two greatest controlling powers of the human system—will and logic.

"*The hand denotes the superior animal,*" said D'Arpentigny, "*the thumb individualizes the* MAN." [100]

¶ 183. D'Arpentigny.

The thumb is divided into three parts—the root [or Mount of Venus], which will be considered fully in a future chapter [*vide* p. 224], belonging more especially to cheiromancy pure and simple; the second phalanx, which is that of Logic; and the first [or nailed] phalanx, which is the seat of the Will. Thus it betrays the whole hand, and interprets the direction in which its indicated aptitudes have been, or are being, developed, for Will, Reason, and Passion are the three prevailing motors of the human race.

¶ 184. The divisions of the thumb.

The second phalanx indicates our greater or less amount of perception, judgment, and reasoning power; the first by its greater or less development indicates the strength of our will, our decision, and our capacity for taking the initiative.

¶ 185. The phalanx of logic and of will.

If the first phalanx is poor, weak, and short, it betrays feebleness of will, want of decision and promptitude in action, unreliability and inconstancy, readiness to accept other people's opinions rather than to act upon one's own, doubt, uncertainty, and indifference.

¶ 186. The upper phalanx weak

When a subject has such a thumb as this, and is at the same time devoted to any particular person or cause, or heroic in his action on any particular emergency, his devotion and heroism are spontaneous and

¶ 187. Heroism of a weak thumb.

---

[100] "*L'Animal supérieur* est dans la main ; *l'homme* est dans le *pouce.*" D'ARPENTIGNY, "La science de la Main ; ou l'Art de reconnaître les tendances de l'intelligence d'après les formes de la main" (Paris, *Dentu*, 1865, 3rd edition).

sudden, [*i.e.*, they are emotional,] not premeditated or lasting.

¶ 188.
Weak will and strong logic.

If with a weak phalanx of will, such as this, your second phalanx [that of reason and logic] is highly developed, you will be able to give excellent reasons for this want of will and uncertainty of disposition, and, though your reasoning powers are excellent, and the promptings of your common sense are strong, you lack the will and decision to put your common sense into practice, and to act boldly on the suggestions of your better judgment.

¶ 189.
The upper phalanx strong, and the lower short.

And conversely, if your first phalanx be long, and your second phalanx be short, you will be quick, impulsive, decided, tenacious of your own opinions [however erroneous they may be], and enthusiastic; but your own want of logic to subdue and direct your spirit of action and strength of will, renders that will of little use to you, and in point of fact you tend towards unreasoning obstinacy.

¶ 190.
Power of strong will.

A well-developed phalanx of Will will often overcome [or at any rate greatly modify] a bad fatality foreshadowed in the palm of the hand.

¶ 191.
Strong will with square fingers, etc.

With soft hand.

With square fingers, and a good line of Apollo [*vide infrâ*, p. 261], a well-developed first phalanx of the thumb indicates a strong will tempered and modified by a love of justice, and with a soft hand this decision of character will only be exercised by fits and starts, in consequence of the natural laziness of the disposition.

¶ 192.
Effect of the Moon.

With a highly-developed Mount of the Moon, a love of repose and quietude will soothe the activity of a highly-developed phalanx of Will, which under these circumstances will only show itself by a dictatorial tone in conversation and a domination in manner.

¶ 193.
Upper phalanx broad.

If the phalanx is broad, but not particularly long, it betrays obstinacy and unreasonableness, unless with square fingers, when it indicates firmness of judgment and the principles and practice of justice.

If besides being the longer the phalanx of will is excessively broad, even to ugliness, it betrays ungovernable passions and obedience to the promptings of an unreasonable will, obstinacy, furious impulse, and exaggeration in all things. Tyrants, murderers, brutal savages, and the like, illustrate greatly this formation, and a man who has this clubbed development of thumb is proportionately to be dreaded as the formation is more or less pronounced. In a hand which is essentially passive this thumb will denote merely morbid melancholy, especially if the phalanx of logic is short, as the latter, if long, will greatly modify the indications of the form.

¶ 194. Excessive broadness of the phalanx of will

In a passive hand.

The sign of the clubbed thumb is, however, the more certain when the Mount and Plain of Mars are high and the line of the head is weak. It may, to a great extent, be modified and corrected by a well-developed Mount of Apollo [vide p. 211], of Jupiter [vide p. 204], or of Venus [vide p. 224], or by a good line of heart. With these modifying signs such a subject will rather injure himself in his fits of temper than wilfully do an injury to another.

¶ 195. Clubbed thumb in a bad hand.

When the phalanx of will turns back, as it often does, it indicates extravagance, luxury, and, with other propitious signs, generosity, though an excess of this formation is bad from its unreasoning unthriftiness, which argues a want of moral sense. If, in addition, the Mounts of Jupiter and Mars are high, the extravagance of the subject will be devoted to display and the gratification of his personal vanity; and, as I have observed before, [¶ 151,] the same remarks apply [though in a lesser degree] to the fingers, which, if turned back, indicate also extravagance.[101]

¶ 196. Upper joint turned back.

[101] Si pollex, qui plerumque extra manum jacet, potius petit mediam palmam mediosque digitos, libidinosam ostendit atque impuram mentem.

¶ 197.
Effect of broad thumb.

It must also be noted that broadness of the first phalanx of the thumb [obstinacy] renders any excess of formation found elsewhere in the hand additionally serious and ominous, for it is almost invariably accompanied by a short and small phalanx of logic or reason.

¶ 198.
Effect of small thumb.

Therefore, it will be seen, that the greater or less development of the various portions of the thumb plays a most important part in the science of cheirosophy: you may take it, as a rule, that a small, ill-formed, feeble, or badly-developed thumb indicates vacillation of mind, irresolution and want of decision in affairs which require to be governed by reason rather than by instinct or by sentiment.

¶ 199.
Short logic in a weak hand.

If the shortness of the second phalanx [logic] shows want of reasoning power; pointed fingers, a weak line of the head declining upon a high Mount of the Moon, and forked at its extremity, all give unfailing indications of a foolish-mindedness that cannot be counteracted even by a well-developed phalanx of will, or a well-formed line of fortune.

¶ 200.
Large thumbs and small thumbs.

Small-thumbed subjects are governed rather by heart, as opposed to large-thumbed subjects, who are governed by head; the former have more sentiments than ideas, the latter have more ideas than sentiments.

¶ 201.
Modifying signs.

The bad indications [*i.e.,* the weakness] of a small thumb may be counteracted by a high Mount or Plain of Mars, which will give firmness and decision to the character, as well as calmness and resignation. Another modifying sign is softness of the hand, [*i.e.,* laziness,] for in this case the subject will not take the trouble to get into mischief, [though he lacks the strength of Will to resist temptation when it comes in his way].

¶ 202.
Effect of large thumb and of small thumb with smooth fingers.

With a large thumb, you will be independent and self-reliant, inclining rather to despotism, governing by will rather than by persuasion; with a small one, you will be reliant on others, easily governed, and wanting in self-confidence, but you will possess, if your fingers be smooth, [*no matter what their termination,*] the instincts, the natural tendencies, [undeveloped though they may be,] of art.

So in the same way he who is poetic or artistic by reason of his smooth, conic fingers, is the more certainly so if he have a small thumb; whilst he who is exact and scientific by reason of his square or knotted fingers, will be the more so if he have also a large thumb.

¶ 203. Small thumb on artistic hand.

Large thumb on scientific hand.

§ 9. *The Consistency of the Hands.*

CONSISTENCY.

Another great class difference which exists among hands is that of consistency. That is to say, of two hands outwardly the same, one may be so firm as to be hard, and the other may be so soft as to be flabby, and the great distinction thus indicated is, that soft hands betray a quiet temperament, inclining to laziness, and reaching even to lethargy, whilst hard hands indicate an energetic longing for action and a love of hard physical or manual labour. These differences show themselves chiefly in the way in which the different subjects undertake their work.

¶ 203a. Indications.

The soft hand has more poetry in its composition than the hard. Thus, an artist with hard hands will paint things real and actual rather than things ideal, and his pictures will be more active and manly than those of a softer-handed artist, who will paint the images of his fancy, and whose works will show greater soul, greater diversity, and more fantasy.

¶ 204. Soft and hard hands. Artists.

Again, a spatulate subject with hard hands will engage in active exercises, athletics, and the like, whilst the similar but softer-handed subject prefers gentler exercise, and prefers to watch others engaging in active occupations; the former will get up early and work hard, whilst the latter will get up later,

¶ 205. Hard and soft spatulate hands

though, when up, he will work as hard, or take great interest in seeing others work as hard.

¶ 206. Soft hands. Again, people with soft hands have always a love of the marvellous, being more nervous, more impressionable, more imaginative than those with hard hands. A *very* soft hand has to a still greater degree developed this fascination for the strange and uncanny, being rendered additionally superstitious by their bodily laziness, which keeps their minds active. The tendency is still more pronounced if the fingers are pointed.

¶ 207. Soft spatulate hands. On the other hand, a *soft* spatulate subject, *by reason of his desire* for movement, is always eager to search and experimentalize in the marvellous; discoveries in the occult sciences are generally made by people with pointed fingers, but these discoveries are always followed up by people with soft spatulate hands.

¶ 208. Very hard hands. In like manner a very hard hand will be superstitious from want of intellect to make him otherwise, and the tendency will be the more accentuated if the subject have also pointed and smooth fingers.

¶ 209. Influence of the thumb. But if a soft hand have a long phalanx of will, the subject, though naturally lazy, will discipline himself, and often compel himself to do work which is distasteful to him.

¶ 210. Hardness of the hands in age. I have in another place [102] called attention to the circumstance that, as we increase in years and our intellects get weaker, we are apt to take to hard manual labour, such as gardening, carpentering, and the like; it will be observed that at the same time our hands get firmer, even to hardness, and this before natural decay renders them parchmenty and bony. We become more philosophic, and less cre-

---

[102] "Chiromancy, or the Science of Palmistry; being a Concise Exposition of the Principles and Practice of the Art of Reading the Hand." By Henry Frith and Edward Heron-Allen. (London. 1883.)

dulous, more logical, and less romantic, as with age our joints thus develope. I have before alluded [*vide suprâ*, ¶ 130] to the fact that joints may develop in a smooth hand, as a result of intellectual and scientific cultivation.

¶ 211. The affections.
Soft hands are often more capable of tenderness and affection than true love; but hard hands are generally the more capable of true love, though less prone to demonstrative tenderness and affection.

¶ 212. The perfect consistency.
To be perfect, a hand should be firm without hardness, and elastic without being flabby; such a hand only hardens very slowly with age, whereas an already very firm hand often becomes extremely hard. Smoothness, and a gentle firmness of the hand, in youth, betoken delicacy of mind, whilst dryness and thinness betray rudeness and insensibility.

¶ 213. Hard hands are like spatulate hands.
A hard hand has, by its hardness, many of the instincts of the spatulate, whatever may be its exterior formation. For instance, it can bear hardships and privations before which a soft-handed subject would succumb. It also likes the life of constant effort and struggle, so distasteful to the soft, and so welcome to the spatulate hand.

¶ 214. Excessive hardness.
It must be also noted that an *exceedingly* hard hand always denotes unintelligence, and if a short phalanx of logic is superadded thereto, the activity of the hand will be ill directed in the pursuit of pleasures and other affairs useless to the owner of the hand.

THE ASPECT.
¶ 215. Wrinkled hands.
The aspect of the hand must also be taken into consideration, in connection with the consistency; thus,—a soft *wrinkled* hand shows impressionability and uprightness of soul, and a wrinkled *hard* hand is that of a person who is pugnacious, irritating, and teasing, especially if the nails be short [*vide* ¶¶ 251-3].

¶ 216. Wrinkled backs.
The *back* of the hand lined and wrinkled always indicates benevolence of mind and sensitiveness of soul.

¶ 217.
Firm hand and lower joint, etc.
A hand of a good firm consistency, having the joint of order [the second] well developed, with a long phalanx of logic, is an almost invariable indication of good fortune, which is well merited, well striven for, and therefore thoroughly realized.

¶ 218.
Soft-handed republicans.
People of sedentary occupations generally have soft hands, and are generally the most republican in their creeds, because their bodies being quiet their brains are the more active. These soft-handed republicans are those who rave at their followers and harangue the mob with the premeditated verbiage of experimental incendiarism, whilst the hard-handed republicans are those who organize, who act, and who devote all their energies to the attainment of the objects which their pointed fingers prompt them to strive for.

¶ 219.
Firm hands and the Mount of Venus.
The man with the firm, strong hands and the developed Mount of Venus [*vide* p. 224] is the man who will exert himself to amuse others with feats of grace and of agility; who will romp with children, and work hard to contribute his share to the general harmony.

¶ 220.
Softness during illness.
During an illness a hand which is naturally inclined to be hard, will often become temporarily soft, regaining its natural hardness when the ordinary habits of life are resumed. It has been argued to me from this, that the indications afforded by cheirognomy are unstable and unreliable; but on the other hand it is a most interesting fact in support of the science, for the enforced laziness during the time of illness produces in the hand the cheirognomical sign of laziness, and proves that a temporary abandonment of its characteristic employments by a hand, will cause it to conform cheirognomically to the indication of the newly-acquired (though enforced) course of life.

§ 10. *The Cheirognomy of the Individual Fingers.*

¶ 221. The first finger.

There is also to be considered a separate cheirognomy of each individual finger, which must particularly be studied in reading the indications of a mixed hand [*vide* p. 170].

Thus, if the first finger [or index] is long, it indicates pride and contemplation; if it is short it indicates activity and impulse; if it is very long [*i.e.*, as long as the second or middle finger] it indicates a sense of luxury even to sensualism, love of pleasure and comfort rather than of art, combined with an indiscriminating arrogance and egoism which is ashamed of poor relations or associates if surprised in their company.

¶ 222. Very long index.

A long and pointed first finger betrays religious exaltation. If it is longer than the second finger, it denotes that the life is ruled by ambition or [if the hand is good in its other developments] by religion.

¶ 223. The phalanges of the first finger.

If the first [or nailed] phalanx is long it denotes religion and intuition; if the second [or middle] phalanx is long it indicates ambition; and if the third [or lowest] phalanx is long it betrays pride and love of domination.

¶ 224. Pointed with a developed mount.

If, whilst of normal length, the finger is pointed, the subject has intuition and religious instincts. If the Mount at the base of the finger [*vide* p. 204] is highly developed, and all the fingers are smooth, we generally find a tendency to ecstacy and mysticism. The intuition of the pointed forefinger applies itself, as a rule, to the contemplation and perfection of the qualities shown by the formations of the other fingers and the rest of the hand.

¶ 225. Square index

If it is square, we find a love of and a search after truth. Such a subject will seek to discover truth from natural [not occult] sources of information. He will have a love of landscape painting in art, whilst, with a good development of the Mount of Jupiter [*vide* p. 204], he will have tolerance and reason in religion.

¶ 226.
Spatulation of the index.

A spatulate termination to this finger [fortunately a very rare form] indicates, as a rule, intense mysticism and error, especially in a smooth-fingered hand.

¶ 227.
The second finger. Spatulated or twisted.

If the second [or middle] finger is highly developed and flat [*i.e.*, inclined to spatulation] it indicates sadness, fatalism, a morbid imagination, and melancholy. [If it is twisted it is said to be a sign of murderous instincts and inclination.]

¶ 228.
Pointed second finger.

This finger is seldom pointed; but when it is so the point modifies the sad and morbid influence which is the inseparable evil of the development and conditions of this finger, producing callousness and frivolity in place of morbidity and moroseness. This result is more striking if the hand bears also a small thumb.

¶ 229.
Square middle finger.

If the finger is square the character of the subject becomes grave in proportion to the greater or lesser accentuation of the square formation of the finger.

¶ 230.
Spatulation of the middle finger.

The spatulate is the most natural and ordinary termination for this finger, giving it activity of imagination, and a morbid fancy in matters relating to art, science, and literature.

¶ 231.
The phalanges of the second finger.

If on this finger the first phalanx is long, it betokens sadness and superstition, *very* long it betrays a morbid desire for death, and, in a weak hand with a small thumb, a horrible temptation to suicide. If the second phalanx is long by comparison with the others it denotes love of agriculture and mechanical occupations, or, if the joints are prominent, mathematics and the exact sciences. If the fingers are smooth the development of this second phalanx will give a talent for occult science. Lastly, if the third phalanx is long and large it denotes avarice.

¶ 232.
Middle finger inclined towards the third or first.

If the finger incline at the tip towards the first finger the fatalism indicated thereby is dominated, and to some extent modified, by pride and self-con-

OF THE INDIVIDUAL FINGERS. 127

fidence. If it incline towards the third [or middle] finger this same fatalism is dominated by art.

If the third [or ring] finger is as long as the first it shows artistic taste, and a desire and ambition to become celebrated and wealthy through artistic talent. If it is as long as the second finger, however, it indicates a gambler, or a person who is foolhardy and rash, especially when the Mount of Mercury [*vide* p. 214] is developed. When a hand is otherwise good and strong, this length of the third finger merely indicates a love of adventure and enterprise, especially if the finger tips are spatulate. If the finger is longer than the second or middle finger, it indicates that the instinct and talent for art will triumph over the fatality which will place obstacles in its way and try to impede its progress. ¶ 233. The third finger

If the tip of this finger is pointed it denotes intuition in art; but if all the other fingers present different formations of the tips it will indicate frivolity and levity of mind. ¶ 234. Pointed third finger.

A square-tipped third finger will seek for positivism, research and reason in art, and, with the third or lowest phalanx large, a love of wealth. ¶ 235. Square third finger.

A spatulate termination to the finger will denote love of action and movement in art, battles, struggles, animated scenes, and representations of them. Such subjects generally make good actors, elocutionists, and orators. ¶ 236. Spatulation of the third finger.

If the finger is amorphic and shapeless at its extremity, it denotes positivism of mind and commercial talent and instinct. If the finger is short whilst the rest of the hand is decidedly artistic, the talent for art will be there, but it will indicate a mercenary pursuit of art for the sake of its emoluments and rewards. ¶ 237. Shapelessness or shortness of the third finger.

The first [or outer] phalanx long shows great artistic feeling; the second, highly developed, denotes reason and industry in art and the love of those ¶ 238. The phalanges of the third finger qualities; the third phalanx dominating the others betrays love of form and conventionality, vanity in art, and a strong desire for wealth.

¶ 239. *The joints of the third finger.* The development of the first [or upper] joint will give to this finger research and love of perfection and finish in art, whilst a prominence of the second [or lower] joint will indicate a love and appreciation of riches.

¶ 240. *The fourth finger.* If the fourth [or little] finger is long [*i.e.*, reaching to the middle of the nailed phalanx of the middle finger] it indicates a search after knowledge, a love of education, and a desire to perfect oneself in all kinds of learning. Such a subject will gather quickly the principia of a science, and [from the eloquence and powers of expression, denoted by a development of this finger] can discourse and converse with ease on any subject he has ever taken up. If the finger is as long as the third itself, the owner of the hand will be a philosopher and a savant, unless the whole hand is bad, when this formation denotes cunning and ruse. In the rare cases where the little finger is so long as to reach the top of the second finger, the indication is that the love of science will dominate every fatality of the life, and will surmount every obstacle which may be thrown in his way. If, on the other hand, the finger is very short, it betokens a very quick perception and power of grasping things and reasoning them out with rapidity.

¶ 241. *Pointed fourth finger.* A pointed little finger indicates intuition in applied and occult sciences, perspicacity, cunning, and eloquence, which can be brought into requisition to discourse about the veriest nothings. Such subjects make by far the best "after-dinner speakers" and complimentary orators.

¶ 242. *Square little finger.* Squareness of this finger tip denotes reason in science, love of research and discovery, combined with logic, good sense, and facility of expression when there is need for it.

A spatulation of the little finger gives movement, agitation, and often fantasy in science, fervid and moving eloquence, with a strong aptitude and talent for mechanics. If the rest of the hand is bad, this spatulated formation of the finger tips will indicate theft.

¶ 243. Spatulation of the little finger

If the first phalanx is long, we find love of science and eloquence; when the second phalanx is the longest of the three, we find industry and commercial capacity; and, with a development of the third, we get cunning, cleverness, perspicacity, and lying.

¶ 244. The phalanges of the fourth finger.

Prominence of the first joint indicates research in science, and often divination; the salience of the second betrays research and industry in business and commercial skill and aptitude.

¶ 245. The joints of the fourth finger.

§ 11. *The Habitual Actions and Natural Positions of the Hands.*

GESTURES.

In arriving at an estimate of a character by the application of cheirosophy, there are also to be considered the habitual actions of the hands and the natural positions into which they unconsciously place themselves when in a state of repose. This branch of the science of Cheirosophy has been treated as a distinct science under the name of Cheirology,[103] and as such I purpose at a future date to treat it. At

¶ 245*. Cheirology.

---

[103] There exists an extremely rare duodecimo by John Bulwer, devoted entirely to this subject, and entitled "Chirologia; or, the Naturall Language of the Hand: Composed of the Speaking Motions and Discoursing Gestures thereof, Whereunto is added, Chironomia: or the Art of Manuall Rhetoricke, Consisting of the Naturall Expressions, digested by Art in the Hand, as the Chiefest Instrument of Eloquence, by Historical Manifestos exemplified, out of the Authentique Registers of Common Life, and Civill Conversation, With Types, or Chyrograms, A longwish'd for illustration of this Argument." By J. B. Gent. Philochirosophus. (London.) Printed by *Thos. Harper*, 1644.

¶ 246.
Closed hands and open hands.

¶ 247.
Carelessness.

¶ 248.
Agitation and quiescence.

¶ 249.
The fingers tapping together.

¶ 250.
Gaule. Gesticulations in speech.

THE FINGER-NAILS.

present, however, a few of the elementary rules of this branch cannot fail to be of use and interest to the student of Cheirosophy, as being in a high degree germane to the considerations wherewith we are in this volume occupying our attention.

To keep the hands always tightly closed denotes secretiveness, and not unfrequently a tendency to untruth. To keep them closed in this manner even when walking betrays timidity and avarice, whilst to carry the hands continually open indicates liberality and openness of disposition.

To let the hands hang carelessly and loosely by the sides betokens laziness, restlessness, and often a suspicious disposition.

If in walking you keep the hands clasped, swinging them to and fro, it shows promptness and impetuosity of character, whilst to keep the hands motionless by the sides betrays dignity and reserve. To keep them absolutely and studiously impassive denotes vanity, conceit, and often falsehood.

If when the body is at rest the fingers are constantly tapping together, it denotes lightness, dreaminess, and fantasy. If they beat together strongly, it indicates promptitude and decision of opinion, whilst, if they tremble, it usually denotes [unless the subject is nervous and highly strung, when it is a natural consequence] folly and often want of principle.

Gaule [*vide* note [97], p. 110] points out the fact that "the often clapping and folding of the handes note covetous, and their much moving in speech loquacious;" two indications which, though correct, partake rather too much of the nature of truisms.

§ 12. *The Finger Nails.*

There existed formerly an extremely ancient art of divination termed Onychomancy, or Divination by the Finger Nails. Into the study of this art much of

THE FINGER NAILS.   131

the charlatanry, superstition, ignorance, and fraud of the *soi-disant* sorcerers of the early and Middle Ages necessarily was infused, and the rites whereby auguries were drawn from rings suspended on the finger nails, or from the figures formed by the reflection of the sun's rays falling upon the finger nails of a child which had been previously polished with oil, are too absurd to receive a moment's consideration when pursuing investigations, the aim of which is the discovery of truth. The finger nails follow of course to a very great extent the shapes of the tips of the fingers, Nature having provided them, not as among the brutes, for purposes of offence and defence, but merely as a protection to the delicate tips of the fingers.[104] Still, very considerable indications of character may be found in the aspect of the finger nails, which, as far as they are interesting to us, in the study of Cheirosophy, are as follows.[105]

¶ 251.
Short nails.

If the nails are short, broad rather than long, with the skin growing far up them, the subject will be pugnacious, critical in disposition, fond of domination and control in matters relating to himself and to his surroundings; in fact, he will be imbued with a spirit of meddlesomeness. His establishment will be minutely ordered, and regularly conducted. With

[104] This is noted by Aristotle in his treatise, ΠΕΡΙ ΣΩΩΝ ΜΟ-ΡΙΩΝ. Βιβλ. Δ' Κεφ. ί. Εὖ δὲ καὶ τὸ τῶν ὀνύχων μεμηχάνηται, τὰ μὲν γὰρ ἄλλα ζῷα ἔχει καὶ πρὸς χρῆσιν αὐτοὺς, τοῖς δ' ἀνθρώποις ἐπικαλυπτήρια σκέπασμα γὰρ τῶν ἀκρωτηρίων εἰσίν.

[105] Democritus Junior, [Robert Burton,] in his "Anatomy of Melancholy" (Oxford, 1621), [Part I., Sect. 2, Memb. 1, Sub-sect. 4,] makes some extremely interesting observations on the subject of the finger nails, which, for the benefit of those who are so far interested in the subject, I here transcribe:—
"Chiromancy hath these aphorisms to foretell melancholy. . . . Baptista Porta makes observations from other parts of the body, as, if a spot be over the spleen, 'or in the nails; if it appear black it signifieth much care, grief, contention, and melancholy'; the reason he refers to the humours, and gives instances in him-

spatulate fingers and a short thumb this subject will be constantly tidying things away, arranging and dusting his rooms for himself, and organizing the dispositions of his property.

¶ 252.
Short-nailed women.

With short nails, a woman whose line of heart is small, whose head line is straight and inclined to turn up towards the little finger, whose Mount of Mercury [*vide* ¶ 468] is flat and covered with lines, and whose Mounts of Moon and of Mars [*vide* pp. 217 and 220] are high, with the joints of the fingers plainly visible, will be undoubtedly of that kind of woman who is known as "a virago." The above are all the signs of harshness and quarrelsomeness in woman, and the possession of short nails accentuates the certainty of the indications.

¶ 253.
Good indications of short nails.

Short nails denote sharpness, quickness of intellect, and perspicacity. With a good line of head they indicate administrative faculty; with a good line of Apollo they indicate irony and *badinage*. Short-nailed subjects make the best journalists, by reason of their love of criticism and readiness to engage in any dispute or contention. On a good-natured and happy hand, or in a lazy hand, short nails denote a spirit of mockery and of good-humoured sarcasm, frivolity, criticism, and contradiction.

<blockquote>
self, that for seven years' space he had such black spots in his nails, and all that while was in perpetual law-suits, controversies for his inheritance, fear, loss of honour, banishment, grief, care, etc., and when his miseries ended the black spots vanished. Cardan, in his book 'De Libris Propriis,' tells such a story of his own person, that a little before his son's death he had a black spot which appeared in one of his nails; and dilated itself as he came nearer to his end. But I am tedious in these toys, which, howsoever in some men's too severe censures they may be held absurd and ridiculous, I am the bolder to insert as not borrowed from circumforean rogues and gypsies, but cut of the writings of worthy philosophers in famous universities, who are able to patronize that which they have said, and vindicate themselves from all cavillers and ignorant persons."
</blockquote>

It goes almost without saying that when the nails are short from the habit of biting them they indicate nervousness, abstraction, subjection to fits of melancholy, a worrying disposition, and continual irritation. This is still more the case if the ends of the fingers are spatulated.

¶ 254. Bitten nails.

White and polished, soft in texture, with a tendency to pinkness by reason of their transparency, and of a normal and well-proportioned length, the nails indicate a good spirit, delicacy of mind, sensitiveness, tact, and good taste.

¶ 255. White smooth nails.

Albertus Magnus states that round and rough nails show a great capacity for attachment, whilst very white or dark-coloured nails betray malignity of disposition.

¶ 256. Extremes of colour and condition.

If the nails are long and curved they denote ferocity and cruelty. According to Gaule, they " signe one brutish, ravenous and unchaste." Short and pale they betoken falsehood and cunning. Black-toned nails are a sign of treachery, and narrow and curled nails of impudence and imposture. With large white nails the morality is good. Round nails are the indications of a luxurious disposition, and very thin nails betray a subtle disposition, generally accompanied by weak health.

¶ 257. Long and curved. Short, dark, narrow, round, and thin.

According to De Peruchio [*op. cit.* p. 116, note [98]] points in the nails, *whether white or black,* mean nothing if they are cloudy and diffused. If they are star-shaped they indicate a vain worship of things to which one is attracted, and if they merely present the appearance of a clearly-marked *point* they are the indications of some event the nature of which is not specified.

¶ 258. Spots in the nails.

White marks upon the thumb-nail denote affection, which is generally reciprocated. Rays of white on the same nail show a useless and ill-directed attachment, whilst black spots denote faults [or even crimes] resulting from passion.

¶ 259. Spots upon the thumb-nails.

¶ 260. **Spots on the finger-nails.** A white mark on the nail of the first, or forefinger, foreshadows a gain, and a black mark a loss. On the nail of the second finger the white mark tells of a voyage, and the black one of impending destruction. On the third finger nail a white mark denotes honour and wealth, and a black mark infamy and baseness. White marks on the little finger nail denote a faith in science and commercial gain. These are the dicta of traditional Onychomancy.

¶ 261. **Sir T. Browne.** Sir Thomas Browne, in his "Pseudoxia Epidemica,"[106] speaks as follows concerning the indications of the finger-nails:—"That temperamental dignotions and conjecture of prevalent humours may be collected from spots in the nails we are not averse to concede; but yet not ready to admit sundry divinations vulgarly raised upon them, nor do we observe it verified in others what **Cardan.** Cardan[107] discovered as a property in himself to have found therein some signes of most events that ever happened to him [*vide* note [106], p. 132]. Or that there is much considerable in that doctrine of Chiromancy, that spots in the top of the nails do signify things past; in the middle things present; and at the bottom things to come. That white specks presage our felicity, blue ones our misfortunes. That those in the nail of the thumb have signification of honour; those in the fore-finger of **Tricasso and Picciolus.** riches; and so respectively in other fingers, as Tricassus,[108] hath taken up, and Picciolus[109] well rejecteth."

[106] Sir Thomas Browne, "Pseudoxia Epidemica: or Enquiries into very many received tenets and commonly presumed Truths" (London, 1646), book v., chap. 24, i.

[107] Girolamo Cardano, "Cardani, de rerum varietate" (Basle, 1557), fol.

[108] Patricio Tricasso, "Chyromantia del Tricasso da Cæsari Mantuano, Nuovame[n]te revista, e con so[m]ma diligc[n]tia corretta" (Venice, 1534), 8vo.

[109] "Antonii Piccioli, seu Rapiti Renovati de manūs inspectione libri tres" (Bergami, 1587).

THE FINGER NAILS.

"To have yellow speckles in the nails of one's hand," says Melton,[110] "is a greate signe of death."

So much, therefore, for the preliminary Cheirognomic examination of the hand generally, and of its various parts and their conditions in particular. It is not necessary to pursue these analytical distinctions further; the student of Cheirognomy will easily understand how to apply the modifications indicated by these combined and analysed indications, to the general tendencies and instincts suggested by the type of a hand so as to arrive at a comprehension of the most accurate *nuances* of the character and constitution of his subject.

¶ 292. Yellow specks

¶ 263. Application of general principles.

[110] JOHN MELTON, "Astrologaster: or the Figure caster" (London, 1620), 8vo.

SUB-SECTION II.

¶ 264.
THE SEVEN
TYPES.

THE SEVEN TYPES OF HANDS, AND THEIR SEVERAL CHARACTERISTICS.

Nomenclature of the Types.

VIEWED by the light of the science of Cheirognomy, all hands belong either to one of six principal classes, or else to a seventh, which is composed of the hands which cannot be rightly classed in any of the other six. These, as determined by M. le Capitaine S. d'Arpentigny, are as follows:—

  I. The Elementary, or Necessary Hand.
  II. The Spatulate, or Active Hand.
  III. The Conical, or Artistic Hand.
  IV. The Square, or Useful Hand.
  V. The Knotty, or Philosophic Hand.
  VI. The Pointed, or Psychic Hand.

Mixed hands.

To these are added [as I have said] a seventh, which is not so much a type by itself as a combination of several. This class comprises those hands which seem to represent more than one type, and are consequently known in Cheirognomy as

  VII. Mixed Hands.

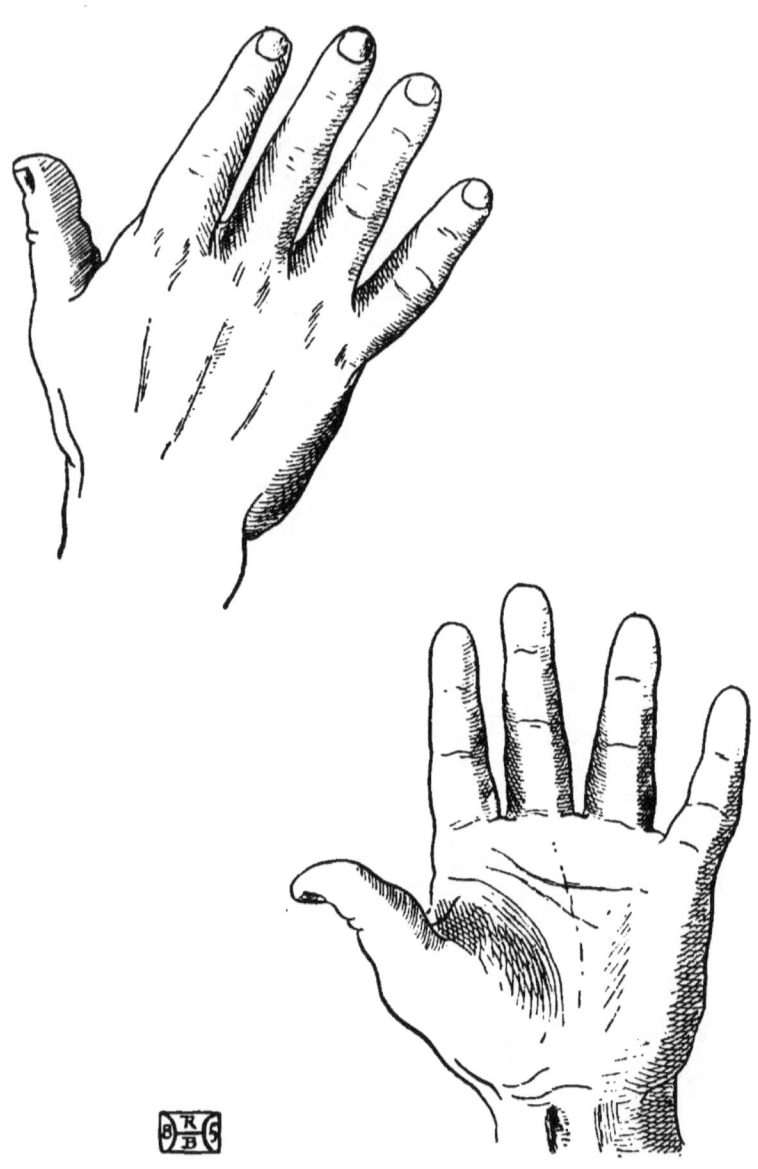

PLATE. I.—THE ELEMENTARY HAND.

§ 1. *The Elementary, or Necessary Hand.*

This is so called because it belongs to the lowest grade of human intelligence, and seems only to be gifted with the amount of intellect requisite to provide the merest necessities of life.

<small>The Elementary Hand.</small>

Its outward appearance presents the following features: the fingers are short and thick, wanting in pliability; the thumb short, often slightly turned back; and the palm very large, thick, and hard. The palm is, as a general rule, longer than the fingers. These hands very frequently have absolutely no line of Fate [or Fortune] at all. [*Vide* Plate I.]

<small>¶ 265. Its appearance and characteristics.</small>

Such a hand as this betokens a crass and sluggard intelligence, incapable of understanding anything but the physical and visible aspect of things, a mind governed by custom and habit, and not by inclination or originality. Such a character, inaccessible to reason from sheer want of originality of intellect to understand it, is sluggish, heavy, and lazy as regards any occupation beyond its accustomed toil. It has no imagination or reasoning power, and will only exert itself mentally or physically so as to obtain that which is absolutely necessary to its existence. Thus in war such hands will only fight to defend themselves, and not for glory or honour; such people fight with a brutish ferocity, but without any attention to the arts of modern warfare. They act by rule and rote, *not* in obedience to their passions or imagination, which are conspicuous by their absence. Such people, having no instinct of cultivation, would regard education as a folly, if not as a crime or as something unholy.

<small>¶ 266. Indications of the type.</small>

The Laplanders are the best specimens of this type; and out of their latitudes the true elementary hand is very seldom found in its pure crass unintel-

<small>¶ 267. Specimens of the type.</small>

ligence, except perhaps among the lower class Tartars and Sclavs, who *exist*, rather *live*, with an existence which is purely negative, dead to any of the higher considerations which make life worth living.

¶ 268. Quasi-elementary hands.
Though, as I say, this type, in its pure state, does not exist among us, still we often see hands with a strong tendency to the elementary form. Such will be noticed amongst mixed hands [*vide infrâ*, p. 170], and it will be found that they always bring these crass and sluggish qualities to interfere with those of the dominant type of the hand.

¶ 269. Powers of cultivation.
Almost the only charm to which these minds are accessible is that of music [and to this I shall recur when considering the mixed hand, *vide* ¶ 346]. Science is an unknown country; they are generally superstitious, and always ignorant; and, having no strength of mind, they are stricken most sorely by any grief or disaster which overtakes them.

¶ 270. Contrasting types.
With the Laps and Sclavs as the examples of this type, we may take the Moslems and Hindoos as the contrasting opposites. Among these poetic, cunning, romantic, sensual peoples the elementary hand does not exist, and to perform the degrading and menial offices which are with us performed by hands showing the developments of this type, these Oriental nations have to employ a separate class of low-caste creatures, to whose elementary hands such labours do not come amiss.

THE SPATULATE HAND.
Its appearance.

§ 2. *The Spatulate, or Active Hand.*

This is the hand whose fingers have the first [or outer] phalanx terminating more or less in a spatule [*vide* Plate II.], and, bearing in mind what I have already said in a previous sub-section [sub.-sec. I., ¶¶ 202-3] concerning the thumb, it will be easily conceived that this latter must be large to give the true character to the spatulate hand.

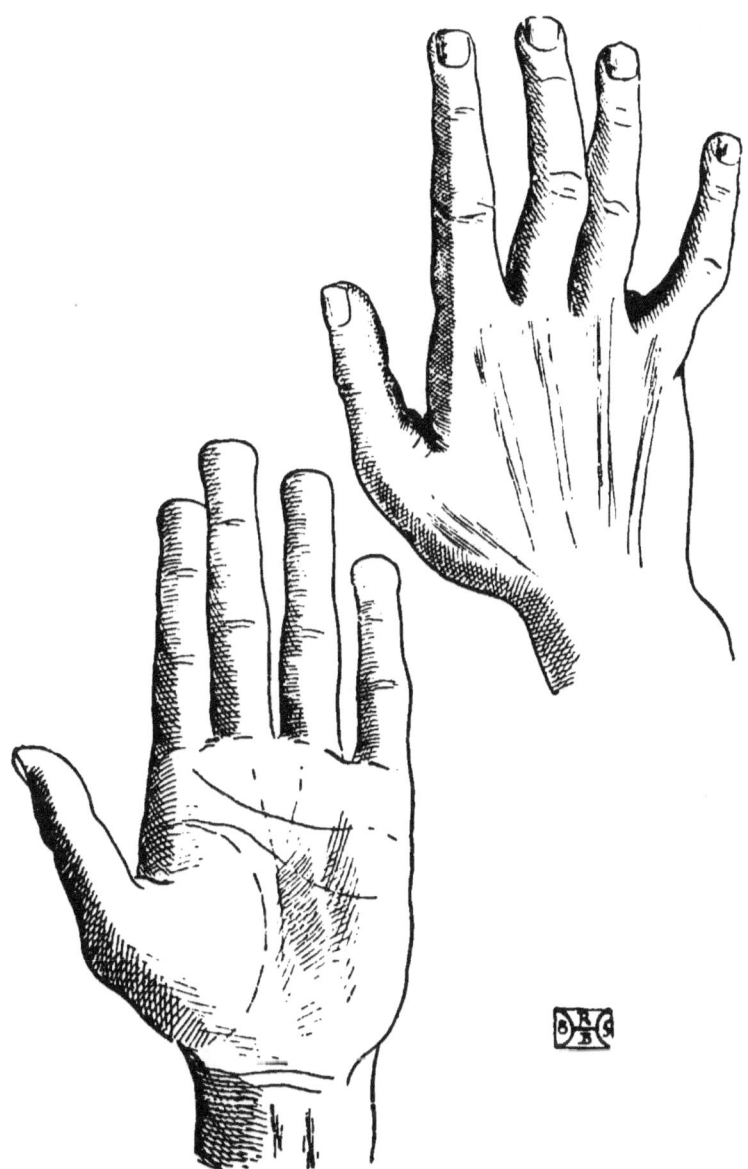

PLATE II.—THE SPATULATE, OR ACTIVE HAND.

The great pronounced characteristics of this type are: action, movement, energy; and, of course, the harder or firmer the hand, the more pronounced will these characteristics be [*vide* ¶ 203]. A man of this type is resolute, self-confident, and desirous of abundance rather than of sufficiency; [it is here that lies the great distinction between the spatulate and elementary hands; the former loving and seeking abundance, the latter requiring only sufficiency;] he will be more active than delicate, more energetic than enthusiastic; in love he will be more constant and faithful [though less tender and affectionate] than the conic or pointed-handed subject, by reason of his want of inclination towards things romantic and poetic.

¶ 271. Indications of the type.

With a small thumb a spatulate subject will try to do much, but will fail, through want of perseverance, to carry out his intentions, from uncertainty in his course of action. He will voyage, but his voyages will be aimless and objectless; he will be active, but his activity will be futile, and produce no results. These diminutions of the force of the type will, however, be greatly modified if the small thumb be largely composed of a long phalanx of logic, [the second,] which formation will reduce within practicable limits his uncertainty, and quicken the intellect to give a better direction to his activity.

¶ 272. Effect of a small thumb.

If, with spatulate tips, his fingers be very smooth he will admire elegance in his surroundings and in the things which conduce to his comfort; but it will be a fashionable rather than an artistic form of elegance. His will be the love of *reality* in art and energy in its pursuit; he will be fond of horses, dogs, navigation, the science of agriculture, the mechanical arts, the theory of warfare, and the talents of administration and command—in fact, all pursuits where the *mind* directs the activity. In all his active pursuits he will seek inspiration for the *motif* of his procedure.

¶ 273. Smooth fingers.

Such subjects are generally musicians, and when this is so, they are great executants. Such subjects, also, are usually self-centred and essentially egotistical.

¶ 274.
Spatulate-handed colonists.
People with spatulate hands make the best colonists, because they are only attached to a country for what it produces for them; they like [*vide* ¶ 271] manual labour and all other forms of activity, being intolerant of insufficiency; if, therefore, their native land is overcrowded, and the good things of this life are scarce, they are quite satisfied to migrate in search of abundance. They are only very slightly sensual, and are greedy rather than epicurean; they like travelling about and seeing new places; being very self-confident, they have no objection to solitude, and are clever at all utilitarian sciences, which enable them to shift for themselves.

¶ 275.
Proclivities of the type.
A man of the spatulate type admires architecture, but likes it to be stupendous rather than ornate. They are great arithmeticians, and to please them things must be astonishing and exact, representing a large amount of physical labour. With them the artizan is more considered than the artist; they appreciate wealth rather than luxury, quantity rather than quality. A town, to suit their views, must be clean, regularly built, substantial, and of business-like appearance.

¶ 276.
Orderliness of the class.
These subjects will be fond of order and regularity because of its appearance, and they will arrange and tidy things more from the desire to be *doing* something than from the love of tidiness itself.

¶ 277.
Administration of the type.
Their laws are strict and often tyrannical, but always just; and their language is forcible rather than ornate. They are brave, industrious, and persevering; not cast down by trifles, but rather courting difficulties so as to surmount them. They desire to command, and are intolerant of restraint, unless for their individual good. They are most tenacious of what is

their own, and are always ready to fight for their rights. Communists, members of exclusive sects and secret societies, and thorough freemasons, are generally of the spatulate type.

¶ 278. Hand of the hereditary nobility.

People who boast of an ancient lineage, and descent from the feudal barons of the Middle Ages, and show in support of their pretensions a fine, pointed, smooth hand, make a great mistake, for the true old stock of the fighting *ancienne noblesse* are always distinguishable by their spatulate fingers. The former must seek a more poetic and romantic origin for their ancestors and descendants.

¶ 279. Occupations of the type.

If the spatulate hand has no need to fight, it will hunt, shoot, fence, race, and, in fact, do anything which conveys the impression of, and satisfies their *penchant* towards activity and strife.

¶ 280. Religion of the type.

In religion the spatulate subject desires a belief reasoned out and certain. His is the domain of Protestantism as opposed to the more *spirituelle*, impressive, and romantic Roman Catholicism. It is for this reason that the northern nations [among whom the spatulate and square types are the most common] are more governed by Protestantism than the conic and pointed-fingered southern races, whose warm, impulsive natures draw them, and make them cling closely to the mysticism and poetry of the Catholic religion. It is for this reason [*i.e.*, their practicality as opposed to romanticism] that in mechanical arts and sciences the Protestant [or northern] nations excel the Catholic [or southern] nations, whose pre-eminence is to be found in the domains of imagination and the fine arts.[111] The *sentiment* of religion is essentially Catholic,

---

[111] " Partout les Protestants, non pas à cause de leur culte mais à cause de leur organisation, surpassent les Catholiques dans les arts mécaniques et sont surpassés par eux dans les arts libéraux. Plus travailleurs, ils sont moins résignés."—D'ARPENTIGNY, " La Science de la Main." (Paris, 1865.) P. 153.

but the *ideas* and *practice* thereof are essentially Protestant.

¶ 281.
Spatulate-handed nations.

The North American is the embodiment of this spatulate type, with his advanced notions, his industry, perseverance, and cunning, his economy, caution, and calculation; and as a result of many of these characteristics, we find the type largely represented in Scotland, far more generally indeed than in England, as a moment's consideration will prove to be natural. [M. d'Arpentigny devotes a complete chapter of his work (*vide note* [111], p. 145) to the consideration of the English hand, during the course of which he presents his readers with a highly interesting and very carefully-worked-out analysis of the English character, as demonstrated by the prevailing characteristics of the English hand, the mistakes of which analysis one loses sight of in the general shrewdness of the essay.]

¶ 282.
Talents of the type.

It is to the spatulate type, therefore, that we owe nearly all our great men in the world of physical exertion, of active enterprise, and of applied science; their watchwords being, from first to last, energy, movement, hardihood, and perseverance.

¶ 283.
Excess of the formation.

The excess of this type [*i.e.*, a too highly developed spatulation of the finger tips] will produce a tyrannical desire for action, and a tendency to be constantly worrying and urging other people to increased activity. Such subjects are constantly finding fault, and their freedom of manner and liberty of thought and expression know no bounds. This excess will also give brusquerie and roughness of manner, especially when the line of life is thin and red; but a good line of heart and well-developed Mount of Venus will reduce these significations to those of a rough good-nature.

¶ 284.
Perversion of the success of the type.

When a hand whose spatulate development is thus in excess has the joints developed and a small thumb, the indication will be that of unsuccessfulness in research and invention, arising from the fact that an excessive activity is perverted by want of will to keep it in check.

If this type of hand have the first joint developed, its owner will be endowed with reasoning faculty and independence of rule in his active pursuits. He will be eminently sceptical of tenderness or affection until its existence is proved to him, intolerant of fanaticism, and dead to the charms of imagination and the interests of eccentricity. His will be the talent of politics; he will object to anything uncomfortable or uncertain; he will hate poetry and enthusiasm, and will be endowed with an extreme self-confidence. This development will give him a spirit of cohesion to his fellow-men, resistance against innovation, and a love of political freedom of the masses.

¶ 285. Spatulate tips and upper joint

With *both* joints developed he will combine with his physical energy exact sciences and practical studies; he will devote himself to all mechanical and constructive arts, navigation, geometry, and the like; he will affect particularly the sciences which regulate the laws of motion or action. Such men make the best inventors and engineers, for the activity of their bodies puts into execution and carries out the suggestions and discoveries of their minds.

¶ 286. Both joints developed.

When a spatulate hand is very soft the spirit of action will have a powerful enemy in an innate laziness [*vide* ¶¶ 206-7]. Such a subject will be a late riser, and a man of sedentary habits; but will love the spectacle and noise of action and movement. He will like to travel and hear about travels, but he will travel comfortably, preferring to hear and read about the actions and movements of others than to be active and energetic himself.

¶ 287. Soft spatulate hand.

A subject whose soft spatulate hands have the first [or upper] joint developed will be constantly forming plans and projects, which will, however, come to

¶ 288. Soft hand with upper joint developed.

nothing unless the phalanx of will is long in the thumb, in which case he will *make* himself carry out his plans [unless, of course, he can get some one else to carry them out for him].

THE CONIC HAND.

§ 3. *The Conical, or Artistic Hand.*

This hand is, in its appearance and in the characteristics of the type which it represents, a great contrast to the one whose consideration we have just relinquished.

¶ 289.
Its three variations.

It is subject also to three variations of formation and concomitant characteristics which modify the indications of the type as regards the ends to which it works. Firstly, a supple hand with a small thumb and a developed though still medium palm. This hand is drawn invariably to what is actually beautiful in art. Secondly, a large hand, rather thick and short, with a large thumb. This hand is endued with a desire of wealth, grandeur, and good fortune. And, thirdly, a large and very firm hand, the palm highly developed. This formation indicates a strong tendency to sensuality. All three are governed by inspiration, and are absolutely unfit for physical and mechanical pursuits; but the first goes into a scheme enthusiastically, the second cunningly, and the third with an aim towards self-gratification.

¶ 290.
Its appearance.

Hands of this type always present the following form [modified, of course, by the conditions enumerated in the last paragraph]: the fingers, slightly broad and large at the third [or lowest] phalanx, grow gradually thinner, till the tips of the first [or nailed] phalanges terminate in a cone [as in Plate III.]. The thumb is generally small and the palm fairly developed.

¶ 291.
Proclivities and indications of the type.

Such a subject will be ruled by impulse and instinct rather than by reason or calculation, and will always be attracted at once by the beautiful aspects

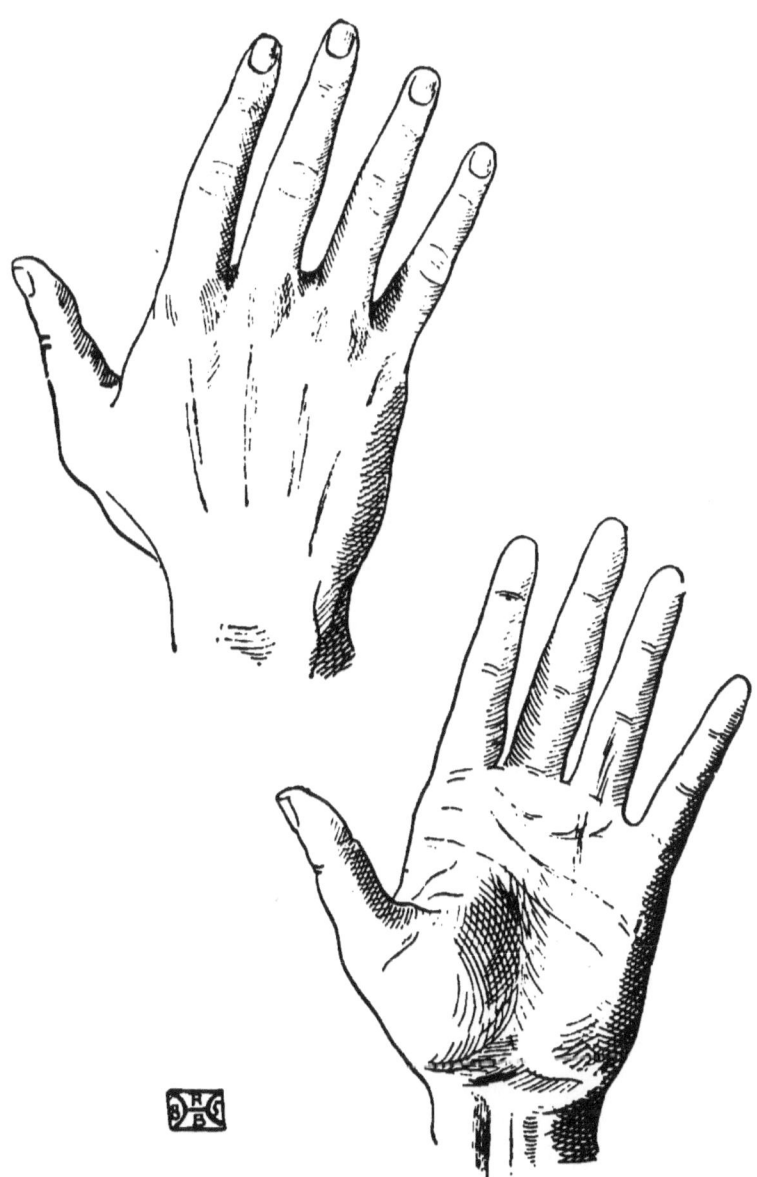

PLATE III.—THE CONIC, OR ARTISTIC HAND.

of life and matter. He will prefer that things should be beautiful rather than that they should be useful. Attracted by ease, novelty, liberty, and anything which strikes his mind as being pleasant, he is at the same time vain, and fearful of ridicule; enthusiastic, but outwardly humble, and his prime motive powers are enthusiasm and impulse, rather than force or determination. Subject to the most sudden changes of temperament, he is at one moment in the seventh heaven of excitable hopefulness, and the next in the nethermost abyss of intangible despair. Unable to command, he is incapable of obedience. He may be attracted in a given direction, but never driven. The ties of a domestic life are unbearable to him. At heart he is a pure Bohemian. In lieu of ideas he has sentiments. Light-hearted, open-handed, and impulsive, his imagination is as warm as his heart is by nature cold. In speech he gesticulates and seeks to impress his meaning by movements of the hands, and he generally succeeds in imparting his enthusiasm to those around him. It is a hard-surfaced hand of this type which characterises the general whose soldiers follow him blindly, who acts on impulse and under excitement for honour and glory, and who leads his men without fear to death or to victory.

¶ 292. Accentuation of the type.

If the characteristics of his type are still more developed [*i.e.*, the palm larger, the fingers smoother and more supple; a small thumb, and the finger tips a more accentuated cone] he is still more the slave of his passions, and he has still less power to hold himself in check. His whole character may be denominated *spirituel*. To him pleasure is a passion, beauty a worship. If he takes up any pursuit he is wild over it. If he makes a friendship, it is an adoration. Never taking the trouble to hate, he never makes enemies. Generous and open-hearted even to extravagance, his purse, which is closed hermetically to his creditors, is always at the service of his friends. He is most sensitive to blame or suspicion, and greatly touched by friendship and kindness. Such subjects will conform to law [so long as it does not interfere with them], because they cannot take the trouble to rebel against it; but they will not brook political despotism which interferes with their comfort, in which cases they will rush enthusiastically to the extremes of republicanism, socialism, and nihilism.

¶ 293. Evil tendencies of the type. Very often in an artistic nature I have found only the defects of the type, sensuality, laziness, egotism, eccentricity, cynicism, dissipation, incapacity for concentration, cunning, falsehood, and exaggeration—a formidable list truly, but a moment's thought will show how easily they may become the besetting sins of an artistic nature. In these cases the hands are large and very firm, the palm highly developed, the Mount of Venus high, and the third [or lowest] phalanx of the fingers always thick and large.

¶ 294. Affections of the type. Subjects of the artistic type are not nearly so capable of constancy in love as their square or spatulate brethren and sisters [*vide supra*, ¶ 271], for they are so apt to fall in love on impulse, and without consideration, whereas with the spatulate, true love, [as are all other subjects,] is a matter of reason and calculation. Again, subjects of the artistic type are, to a great extent, incapable of warm, platonic affection,—filial, paternal, or otherwise,—for in all their emotions they seek the pleasure of the senses rather than the mental and moral satisfactions of attachment.

¶ 295. The characteristics of the type. Beauty is the guiding principle of these hands, but were the world to be entirely populated by them, want of foresight, folly, splendid poverty, and the fanaticism of form would be universal. The artistic type may, therefore, be summed up thus; its prevailing characteristics are love of the beautiful, pre-

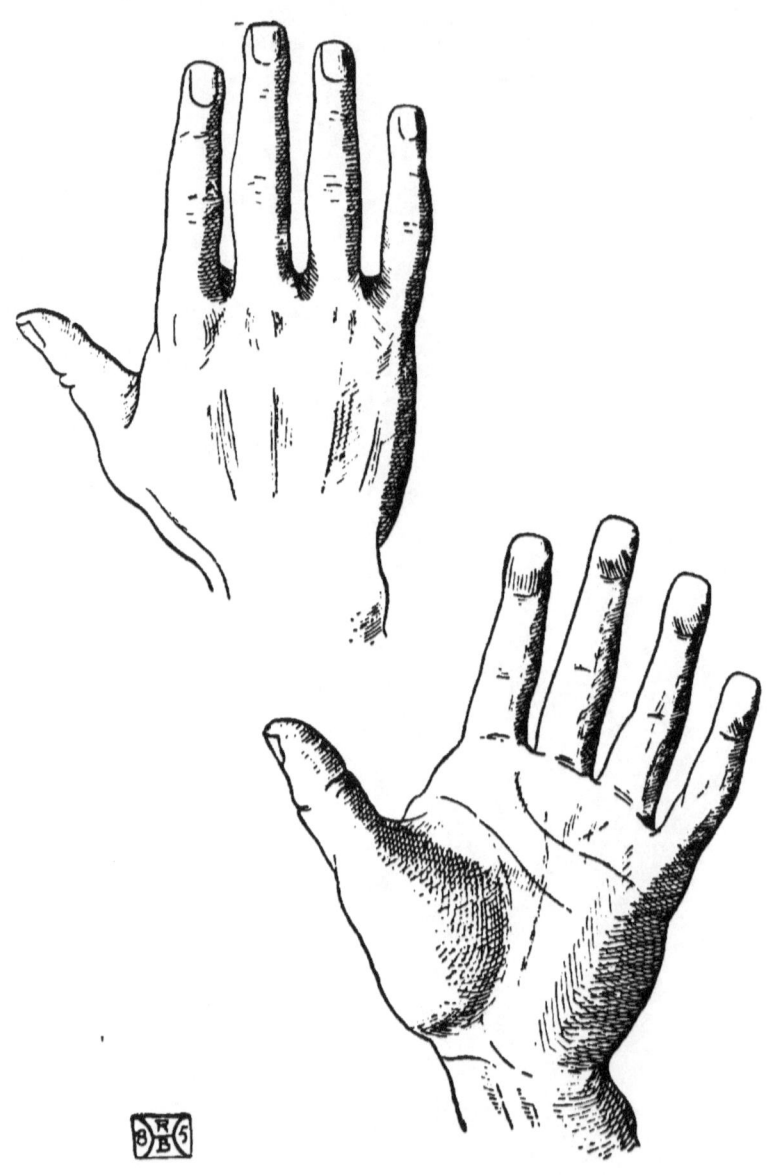

PLATE IV.—THE SQUARE, OR USEFUL HAND.

THE SQUARE, OR USEFUL HAND. 155

ference of the ideal to the real, intuition, impulse, and egotism; and its motto is, " En tout cherchez l'amiable."

§ 4. *The Square, or Useful Hand.*

<small>THE SQUARE TYPE.</small>

This hand generally inclines to size rather than to smallness, the size being usually produced by an increased breadth of the hand, the fingers knotted [*i.e.*, with one or both joints developed, generally, in fact, nearly always, the second or lower one], the outer phalanx square [*i.e.*, the fingers throughout their length having four distinct sides, not being rounded, as is the case with an artistic or psychic hand], the thumb rather large, with the root [Mount of Venus, *vide* p. 224] well developed; the palm of a medium thickness, hollow, and rather firm [*vide Plate* IV.].

<small>¶ 296. Its appearance.</small>

The leading instincts on which this hand founds all its characteristics are perseverance, foresight, order, and regularity. To these hands the useful is far preferable to the beautiful; their great passion is organisation, arrangement, classification, regularity of form and outline, and the acceptation of things prescribed and understood as customary. They like things of a sort to match, and they have essentially the talent of perceiving in things apparently different the points of similarity, and *per contrâ* in things outwardly similar the points of difference. They are great disciplinarians, preferring the good of the community to the welfare of the individual. They are only romantic within the bounds of reason, and are constant in love, more from a sense of the fitness of things than from depth of feeling. They have the greatest aptitude for conforming to the observances of social life, for they are great respecters of persons, and submissive to established authority, from their great love of regularity and order in human affairs.

<small>¶ 297. Indications of the type.</small>

¶ 298.
Square and spatulate types compared.

We find the same submission to authority in the character of subjects of the spatulate type, but with them it arises from another cause. The spatulate subject submits from personal love of his superior, to whom he naturally attaches himself, whilst the square-handed subject submits from admiration of the principles of constituted authority. The dictator must be *powerful* to obtain the allegiance of the spatulate subject; he need only be *properly constituted* to be sure of the allegiance of the square.

¶ 299.
Proclivities of the type.

They cherish their privileges, preferring them to complete liberty; and they have a passion for varied experience, which they are always ready to pay for, preferring acquired knowledge to intuitive perception.

¶ 300.
Religion.

A *Croix mystique* [*vide* ¶ 696] in a square hand will give it calm and reasonable religion.

¶ 301.
Orderliness of the type.

They are slaves to arrangement—that is, they have a place for everything, and everything is in its place; unless their fingers have also the joints developed, it is quite possible [if not probable] that their rooms and cupboards may be outwardly very untidy, but, nevertheless, they always know where everything is. Their books, of which they keep catalogues and indexes, are inscribed with their names and the date of acquisition, and are arranged more in subjects than in sizes, though they love to see them in even sizes as much as possible. They are natty and handy with their fingers, neat and well brushed in their persons, polite and courteous in their manners, whilst they are great sticklers for the ordinances of etiquette.

¶ 302.
Indications of the type.

As a rule, they will only comprehend things as far as they can positively see them, having themselves far too well under control to allow themselves to launch into enthusiasm; they are, therefore, strong disciplinarians, prone to details, fond of minutiæ. Their course of life is regular and pre-arranged, they are punctual, and intolerant of unpunctuality, except when they can regard it as a foil for their own exactness; for they are always vain even to conceit, though they are always too well bred to obtrude their vanity in its more usual and vulgar forms. They are graceful in their movements, generally good shots, and good at games and exercises of skill, as opposed to exercises of mere physical strength.

The best musicians [especially harmonists and musical theoreticians] have always delicately squared fingers, with slightly developed joints and small thumbs [*vide* ¶ 807].

¶ 303. Musicians of the type.

Square-handed people can always govern the expressions of their faces, their language, and their looks; they are most averse to sudden changes of temperament or circumstance. Moderate, *rangés*, they mistake the perfect for the beautiful; they cannot bear excitements and "scenes," and they hate when people obtrude their troubles, discomforts, or quarrels upon them. They dress very quietly, but always very well, and they avoid studiously anything like ostentation, or display in matters of eccentricity, ornament, or jewellery, excepting on fitting occasions, when their magnificence is striking from its good taste.

¶ 304. Manners of the type.

They like poetry to be neat and geometrically perfect, rather than grand or rugged; they call things by generic rather than by specific or distinctive names, and prefer terms which express the use of a thing rather than its appearance. They are generally suspicious and quietly cunning, vigilant, and complete masters of intrigue; they prefer common sense to genius, and social observance to either; they are often flatterers, and are themselves most susceptible to flattery, ambitious, but quietly and steadily rather than enthusiastically and obviously so. They worship

¶ 305. Further characteristics of the type.

talent and cultivation, though without sycophancy; they are fond of arithmetical calculations, though very often not clever at them themselves, unless their thumbs be large, and the Line of Apollo [*vide* p. 261] absent, which are signs which always betray a talent for mathematics. They are good talkers, listeners, and entertainers; they make many acquaintances, but few friends. They do not require men to be sociable so much as blind to the faults of themselves and others.

¶ 306.
Smooth square fingers.

When the fingers, besides being square, are decidedly smooth, the subject will take poetical views of things material and useful, and will affect the study of moral sciences, philosophy, metaphysics, and the like. He will have the instincts of art, and require truth therein: in poetry he will require rhythm, form, and period. Such a mind is well regulated, and he will check a natural tendency to enthusiasm.

¶ 307.
With declining line of head.

The smooth, square hand, is one of the cleverest that exist; love of *truth* in matters which concern itself is one of its first principles, *but* if the line of head come down upon the Mount of the Moon [*vide infrâ*, ¶ 583], this instinct will often be annulled, especially if the line is forked, but there will always be an order and a method in the chimæras to which such a subject is irresistibly addicted, which gives them a strong semblance of truth.

¶ 308.
Upper joint developed.

A square hand, if it has the first joint developed, will have the great advantage over its fellows of the type, of a sincerity, a love of progress and justice which elevates it above the defects of its class. Its calm and cool research after truth will cause it to require reason in matters of art, and to object to anything *outré* or unaccustomed. Law and rule are the necessities of its life.

¶ 309.
Both joints developed.

If both joints are developed, it will indicate a great love of elegant sciences, of the studies of botany, archæology, history, law, and orthography, geometry, grammar, mathematics, and agriculture. This subject will be aggressively methodical, and will insist upon ticketing, docketing, classifying, arranging everything and upon doing everything according to rule, or to a prearranged order.

¶ 310. With Mount of the Moon.

He will be fond of clearly-defined and ascertained studies. [History and politics rather than metaphysics or occult science.] But a small thumb, or a high Mount of the Moon, will give such a subject as this a strange faculty for occultism. At the same time he will have a strongly developed instinct of justice, and is thoroughly trustworthy and true.

¶ 311. Character of the type.

Good sense, therefore, is the guiding principle of the square type, but, were the world wholly populated by them, fanatical "red-tapeism" and narrow-minded despotism would be universal.

¶ 312. Excess of the type.

Excess of this formation will give fanaticism of order and method, despotism in discipline, and narrow-mindedness.

¶ 313. The finger-nails.

A square-fingered hand, to be perfect, should have short nails [argument and self-defence] to defend its love of justice.

It would be easy to continue the interesting subject of the square type to a considerable length if space would allow of it, but we must leave these precise but insincere hands, to turn our attention to another type, which in some respects resembles them, and this is:—

§ 5. *The Knotty, or Philosophic Hand.*

THE KNOTTY TYPE.

M. d'Arpentigny divides this type into two classes or sections: one, that of the sensualists, whose ideas are derived from external influences; and the other that of the idealists, whose ideas are evolved from their inner consciousness.

¶ 314. Its divisions

¶ 315.
Its appearance.
The appearance of the hands of this type is most distinctive. A large elastic palm, both joints developed, the outer phalanx presenting the mixed appearance of the square and of the conic finger tips. This formation, combined with the development of the first or upper joint, gives the finger tips an oval, clubbed appearance, which is rather ugly, but very characteristic. The thumb is always large, having its two phalanges [those of will and of logic] of exactly the same length, indicating a balancing proportion of will and common sense [*vide* Plate V.].

¶ 316.
Its characteristics.
The great characteristics indicated by this type of hand are—analysis, meditation, philosophy, deduction, poetry of *reason*, independence, often deism and democracy, and the search after, and love of, the abstract and absolute truth. The development of the joints gives this hand calculation, method, and deduction; the quasi-conic formation of the exterior phalanx gives it the instinct of poetry in the abstract, and beauty in things real; and the thumb gives it perseverance in its metaphysical studies. In all things these subjects desire truth more than beauty, and prefer the meaning of a sentiment to the manner in which it is expressed: thus, their literature is remarkable for its clearness, its utility, and its variety as opposed to that of the square type, which [*vide suprâ*, ¶¶ 305-6] is notable principally for its finish and regularity of style, and they are great lovers and students of the pure sciences—whether moral, physical, æsthetic, or experimental.

¶ 317.
Its religion.
Such subjects like to account for everything, to know the reason of everything, whether physical, metaphysical, physiological, or psychic; their ideas they form for themselves, without caring in the least for those of other people; their convictions—religious, social, and otherwise—are only acquired as the result

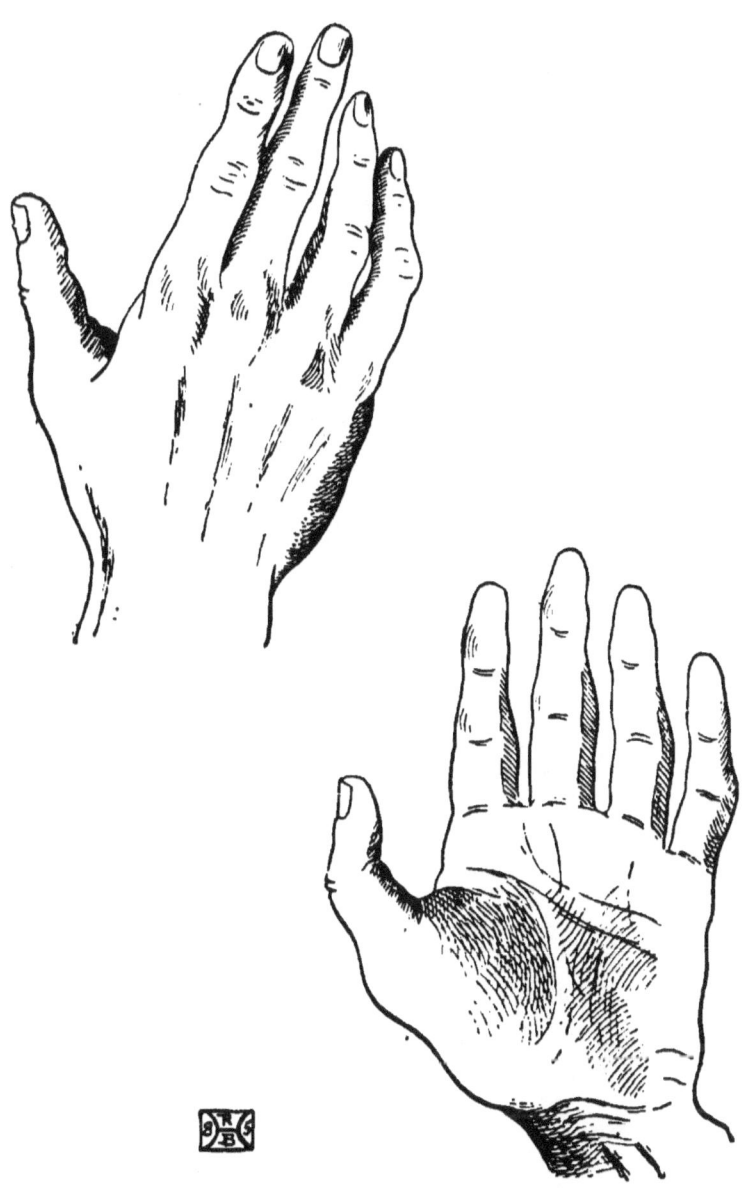

PLATE V.—THE KNOTTY, OR PHILOSOPHIC HAND.

THE KNOTTY, OR PHILOSOPHIC HAND. 163

of careful analysis and consideration of the questions involved; love, instinct, faith, are all made subordinate to reason, which is the principle more powerful with them than rule, conventionalism, inclination, or love, except in matters of religion, for their religion is one rather of love and adoration than of fear and conventionality. It is thus that among the subjects of this type we find a large proportion of persons who become known as sceptics of various kinds, for they look upon doubt and scepticism as one of the first necessary evils of life, which will give way to reverence and adoration, and therefore do not in any way worry themselves on this account.

The subjects of the philosophic type do not study detail to the exclusion of entirety, or the individual to the exclusion of the community, but are capable of considering and comprehending the synthesis *and* the analysis of any subject to which they may turn their attention. Therefore, they are tolerant of all forms of rule, seeing at once the good and the bad points of any or every system of government. ¶ 318. Synthesis and analysis.

They are just, [from an intuitive sense of justice and a discriminating instinct of ethics,] unsuperstitious, great advocates of social and religious freedom, and moderate in their pleasures. It is in these respects that they differ so totally from the subordination and conventionalism of square-fingered hands. ¶ 319. Further characteristics

Thus reasoning out everything, the philosophic type constitute almost entirely the vast schools of the Eclectics. And besides hands which are distinctly of this philosophic type other types may, by the development of joints, attain [as we have seen] attributes of this one. Thus: a square [or useful] or a spatulate [or active] hand may have its joints developed; this will give them [as we have seen] a love of theorizing and speculating on matters of practice, reality, and custom. In the same way a ¶ 320. Philosophic development of other types

conic [or artistic] hand, whose joints are developed, will search after truth in matters appertaining to art, and will speculate upon, and analyze the means of attaining the *truly* beautiful.

¶ 321.
Small and large hands.

If the philosophic hand is small, it thinks and reasons from the heart, studying the entireties of matters which present themselves in masses; if large, and with a proportionate thumb, it thinks and reasons with the head, studying the analysis of those masses, but the result is always the same.

¶ 322.
The guiding principle.

Attained possibly by different means, the end is always identical, "En tout cherchez le *vrai*," and in all things directed by reason, and by common sense directed by will. *Reason* is the guiding principle of these hands, but, were the world wholly populated by them, fanatical reasoning and unregulated doubt and liberty of opinions would be universal.

The Pointed Type.

§ 6. *The Pointed, or Psychic Hand.*

¶ 323.
Its rarity.

We have now reached the consideration of the most beautiful and delicate, but, alas! the most useless and impractical type of hand. This hand is, unfortunately, rare, but when you do see it you cannot help remarking it, and will therefore recognize it at once by its description.

¶ 324.
Its appearance.

It is very small and delicate, having a thin palm, smooth, fine fingers, long and delicately pointed, [as in Plate VI.] or with its joints only just indicated by a very slight swelling. It has generally a pretty little thumb.

¶ 325.
Their guiding principle.

To these subjects belong the domains of the beautiful ideal, the land of dreams, of Utopian ideas, and of artistic fervour; they have the delicacy and true instinct of art of the conic hand, without its bad points, its sensualism, its egotism, and its worldliness. They are guided only by their idealism, by impulse, by their instinct of right in the abstract, and by their

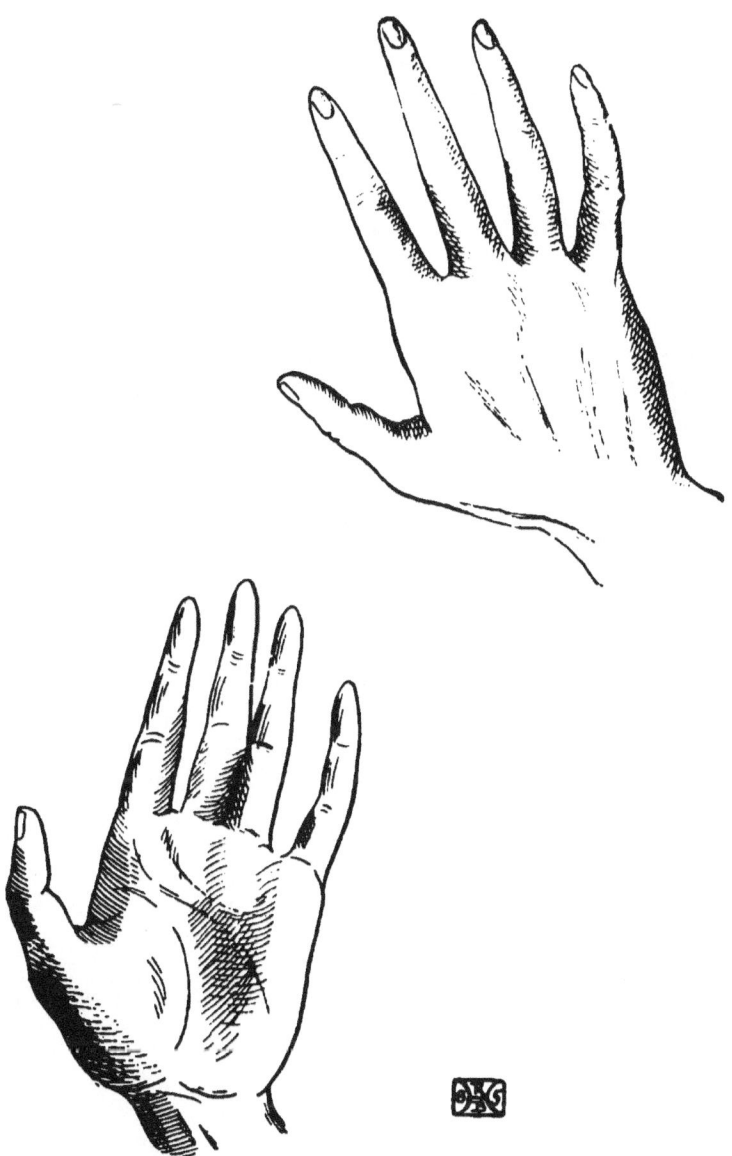

PLATE VI.—THE POINTED, OR PSYCHIC HAND.

natural love and attraction for the beautiful in all things, whether mundane or celestial; bearing the same relation to the philosophic hands that the artistic bear to the useful, the relation of contrast.

These hands never command, for they establish for themselves far too lofty an ideal to care about earthly domination or material interests of any kind; they are incapable of strife or struggles for glory, but, if their instincts of the ideally just are aroused, they will devote themselves even to death in defence of what they consider to be ethically right. Such were the heroes of La Vendee, such were the persecuted followers of Huz, and in such manner is accounted for the devoted enthusiasm of the Moorish and Moslem tribes, who fight like wild beasts for the defence of their faiths, for Allah, his prophet, and the Qur'ân. They will undertake huge forlorn enterprises, but will disdain to embark upon small practicable expeditions, in quest of some material good.

¶ 326. Their characteristics

Idealism

[Desbarrolles, in calling attention to the fact that they not unfrequently have the gift of prophecy, says that he considers their attributes due to the fact that the absence of joints produces a clear passage for the currents of animal electricity or magnetism to which he considers the gifts of prophecy, presentiment, divination, and even ordinary intuition attributable. He considers that the joints in the fingers of the other types act as obstacles to the passage of "la fluide" [which he believes man to receive (if at all) at his finger tips] to a greater or less degree as those joints are more or less developed, and that to the absence of these joints in the psychic hand their intuition, divination, and presentiments are to be ascribed.]

¶ 327. Desbarrolles, the gift of prophecy

An artist with hands presenting the appearance of a psychic formation will paint subjects of wild

¶ 328. Artists.

romance, but will not seek to paint ideas which convey an impression of *truth*. Such subjects have no instinct of real life, nor are they [as the absence of joints would denote, (*vide* ¶ 118,)] orderly in themselves or in their ideas.

¶ 329. Hereditary hands.
As I have said before, [*vide suprâ*, ¶ 278,] these hands are *not* the exclusive inheritance of noble birth; we find them in all classes of life from the highest to the lowest, and wherever and whenever they are found their characteristics are the same, worldly uselessness, with æsthetic perfection and poetry of soul in their highest state of development. Soul is with them the first consideration; form or treatment is with them subordinate to subject; as is also execution to idea: amongst all classes they are respected for their very incomprehensibility.

¶ 330. Excessive development of the type.
The excess of this type [*i.e.*, when the exaggeration of the pointedness is extremely marked] produces romancers, posturists, fanatics of various kinds, persons prone to fantasies and ecstasies, foolhardiness and deceitfulness, and often mysticism. If the line of heart is strong we find in this case excess of affection, which is carried to an extreme, and affectation of manner.

¶ 331. Orientals and spiritualists.
The luxurious dreaming Orientals are almost exclusively of this type. Among them we find spiritualists, mediums, and all the so-called "weak-minded" devotees of psychical science, who accept all that is told them without investigation or analysis, and are therefore the easy prey of "spiritualistic" impostors. In countries where such hands predominate and hold the reins of government we find that rule is maintained by superstition, by the priests, and by fetichism.

¶ 332. Psychic administration and religion.
These subjects can, however, see beauty and good in every form of rule and government from Autocracy to Republicanism, and in every form of belief from

Popery to Positivism. It is the psychic hand that invents a religion, and it is the philosophic, the useful, and the active hands that dissect that religion, and analyse its claims to consideration.

¶ 333. Sympathy of the type.

Such subjects are ruled by heart and by soul; their feelings are acute, their nerves highly strung, and they are easily fired with a wondrous enthusiasm. Theirs are the talents which produce the most inspired poetry; their influence over the masses is extreme, from their power of communicating their enthusiasm to their fellow-men, a power whereby they appeal alike to the most refined and to the most coarse, to the most intellectual and to the most ignorant.

¶ 334. Value of the type.

I could not in many pages give these beautiful useless hands their due. We can only congratulate ourselves that their refining influence exists among us, and that we of the spatulate and square types can work to support them, instead of allowing the world to crush their beautiful characteristics and dull the keenness of their pure intuition.

¶ 335. Upper joint developed.

If the pointed hand have the first joint developed, the character of the owner of that hand will be changeable, and apt to rush from one extreme to another, from ecstatic enthusiasm to suspicion, scepticism, and levity; he will be essentially credulous in things savouring of the marvellous and the mystic; he will be eccentric, and unable to reconcile himself to any prescribed religion; it is such subjects that become fanatics and religious monomaniacs.

¶ 336. Religion of t type.

They have the inspiration and intuition of truth, with a continual desire to analyse their impulses, and to master their romantic emotions. This often causes them to separate themselves from all recognized forms of belief, and to strike out for themselves new religions to satisfy the romantic instinct of piety, which with them is so strongly developed.

¶ 337.
Both joints developed.

With both joints developed, a psychic hand will lose much of its exaltation of character by mingling it with calculation, reason, positivism, and the faculty *of invention;* at the same time it cannot complete and develop its inventions and calculations itself, but leaves them unfinished for square and spatulate hands to work out.

¶ 338.
Effect of joints.

Such a subject, unless his thumb is large, will be prone to discontentedness, doubt, fear, and dejection; and also, with a weak hand, will be Utopian and revolutionary in his views from his very instinct of calculation. These jointed hands of the psychic type have often all their spirit of spontaneous impulse annulled or levelled; their artistic intuitions are spoilt by their instincts of calculation and invention, but *still,* in that calculation and invention the old inspiration and intuition will make itself felt and apparent.

¶ 339.
Hard hands.

Sometimes a hand of the pointed type is hard. This will betoken an artistic use of strength, as in the case of dancers, jugglers, acrobats, and the like.

THE MIXED TYPES.

§ 7. *The Mixed Hand.*

It is here that the task of the cheirognomist becomes most difficult, calls forth all his intuitive perception and skill of analysis, and gives him the greatest difficulty in putting his perceptions into words.

¶ 340.
Its constitution.

The mixed hand is that one of which the shape is so uncertain, as to resemble, even to possibility of confusion, more than one type. Thus, an artistic hand may be so marked in its conicality as to become almost psychic; a square hand may be confounded with a spatulate; or, having developed joints and a quasi-conic tip, may be mistaken for a philosophic, and so on *ad infinitum.* In such cases the cheirosophist must so combine, mentally, the tendencies of both types represented, as to arrive at a true analysis of the character of the subject under examination.

THE MIXED HAND.

¶ 341. Difficulty of interpretation.

In reading the indications afforded by these mixed hands, you will do well to bear very carefully in mind what I have said in a previous sub-section concerning the cheirognomy of the individual fingers [*vide* S.S. I., § 10].

¶ 342. Value of the type.

But as they are the links between the types, so also they are the necessary links which unite the various classes of society, which, without these connecting spirits, would either remain entirely separated, or would be constantly at variance with one another. To the mixed hand belongs the talent of dealing between people as merchants or administrators of justice. They succeed best in intermediary arts—*i.e.*, those of a plastic, regular, and acquired description, such as illumination, carving, heraldry, or decoration. A man endowed with a mixed hand may generally be described as "Jack of all trades and master of none." Such people are less exclusive, and more tolerant of all classes and creeds than those of the pronounced and certain types.

¶ 343. Characteristics of the type.

Such subjects attain to a certain skill in a quantity of pursuits, but seldom attain to an excellence in, or a complete mastery of, any particular one; they have been well described as handy, interesting men, who, to talk to, are always amusing, but seldom if ever instructive. Their intelligence is large and comprehensive rather than strong in any particular direction; they can suit themselves instantly to the company in which they find themselves, and can generally make themselves at home in any discussion which may arise.

¶ 344. Advisability of selection.

The only chance they have of becoming really distinguished, is to take the best talent they have, and cultivate that one to the exclusion of the others; but they seldom have the strength of purpose to effect this.

¶ 345. Advantages of the type.

At the same time there are cases where it may be an advantage to possess a mixed hand—as, for instance, where the idealism of a pointed hand is modified and subdued to reason by the fusion of the square hand, such a hand combining imagery *and* reason.

¶ 346.
Combination of the artistic and elementary types.
A common form of mixed hand is that which combines the artistic and the elementary; and this becomes more comprehensible if you have followed what I said anent those two types; for, as I have pointed out, the intelligence of art or music, and the worship of the beautiful, are the only feelings to which the true elementary hand is at all susceptible, and the artistic hand, by the exaggeration of its failings, may often degenerate into the artistico-elementary. Such a hand will betoken a vacillating, unreliable, apathetic character, without sympathy for the misfortunes, or gratification at the good luck of others. Such people are rude poets, superstitious, and very sensitive to bodily pain. Such hands denote activity by their hardness, and credulity by their pointed tips.

¶ 347.
Appearance and instincts.
Hands of the artistico-elementary type are softer and narrower than those of the purely elementary variety, their fingers are thick and smooth, the thumb gross and conic, the hands closing more easily than they open. Their prevailing instincts are selfishness and greed; they are not good at manual labour or industry of any sort, but they excel in negotiations and schemes of self-aggrandisement. It is this hand that we find in its highest state of development among the low-class commercial races of the Jews.

¶ 348.
Elementary and psychic.
The ignorant enthusiasm which I attributed to the Vendeans [¶ 326] with their love of ease and repose, is the natural result of a step further in this direction— *i.e.*, the mixture of the elementary with the psychic hand. A hand half psychic and half elementary will give us innocence and want of capacity for self-protection. Such subjects will be constantly deceived by the unprincipled. They have no head for business, but only a desire for a quiet, passive, Arcadian life, unin-

tellectual and absolutely harmless, *until* a poetic idea of justice shall rouse it, when its bigoted enthusiasm is as sublime as it is deplorable.

¶ 349. Square and conic.
A combination of the square and the conic will give us the *finesse* and cunning of the square type, with the demoralization of the conic, and the result will be a great hypocrisy and talent for deception.

¶ 350. Square and spatulate.
If your hand is at the same time square and spatulated, to the energy of the spatulate hand will be added the exactitude, the regularity of the square type. You will have the same love of colossal architecture, but will require it to be regular and arranged. You will have the talents of the tactician, of the strategist, of the diplomatist, and of the constructive scientist. Theory, method, and science will be the mainsprings of your activity.

¶ 351. Square and spatulate.
Squareness, confounding itself with spatulation, will give you a love of the minutiæ of an unintellectual existence. You will love to do your own menial work for yourself. You will have a wonderful *practical* knowledge, which will incline to a fanaticism of admiration for things which are practical and useful.

¶ 352. General characteristics.
And so on. These fusions are practically without limit, and, as I have said, it is the task of the cheirognomist, which of all others brings out his skill and aptitude in the science of cheirosophy, to decipher and properly to interpret them. Their prevailing character is always [as may be supposed] vivacity, ubiquity, plurality of pursuits and accomplishments, combined with laziness, insincerity, and want of application and perseverance.

SUB-SECTION III.

THE FEMALE HAND.

THE CHEIROGNOMY OF THE FEMALE HAND.

¶ 353.
Preliminary

ALL that I have said in the preceding sub-sections must [it should be understood] be taken to apply to woman as well as to man; but at the same time the cheirosophist must take into consideration the vast differences of constitution which exist between the sexes, and which, in fact, constitute the base of the relative positions in which they stand to one another.

For without these differences that perfect combination—that of the male and female mind—could not be formed. The physical energy of the man absolutely requires the passive, instinctive tact of the woman to make his energy of any use to him. Thus the woman originates an idea, but the man carries out its active principles. The man will create, what the woman has imagined. The idea of the woman is generally rough, undefined, and vague; by its passage through the man's brain it becomes clear, defined, and practicable. A man cannot acquire the woman's tact and instinctive intelligence without losing much of his virile power, and in like manner a woman cannot educate her mind to develop the energy and power of the man without losing much of her natural and instinctive talent. It reminds one of the Oriental fable of the camel, who, praying for horns, lost its ears. If men would be content to be guided in matters of principle by clever women, and women would allow themselves to be governed in matters of practice by men, the mutual advantage would be incalculable. Proofs of the truth of this are to be found in the fact that the greatest men who ever lived have had clever wives, whilst in the East, where the women are treated either as beasts of burden or as pet animals, the nations are becoming daily deteriorated, and of less importance, and will gradually die out. Men reign and command, whilst women demand and govern: men make our laws, but women our morals; for though the man is more genuine than the woman, the woman is deeper than the man. They argue with the heart, we with the head: we are sensual where they are sensitive and passionate. Their sentiments and ideas are generally truer than our careful reasonings, and they meet our reflection and comprehension with their more subtle intuition and power of analysis. These preliminary remarks it has seemed necessary to make before entering upon the consideration of this sub-section, and five minutes' reflection on them will greatly assist the student of cheirosophy in applying its principles to the interpretation of the female hand.

Thus, the characteristics of the more powerful types [such as the spatulate and the square] will be much less developed with them than with men, by reason of the greater softness which always characterises the hands of the "softer sex." In like manner only very few women have knotty hands—a circumstance arising from that absence of physical and mental combination and calculation, which, as a rule, characterises their movements. Thus they work more by tact than

¶ 354. Difference in the effects of types.

by knowledge, more by quickness of brain than by rapidity of action, and more by imagination and intuition than by judgment or combination.

¶ 355. Jointed fingers. *When* a woman has knotty fingers, she is less impressionable, less imaginative, less tasteful, less fantastic, and more reasonable.

¶ 356. Effects of the thumb. If a woman have a large thumb, she is more intelligent than intuitively quick. If she have a small thumb, she is quicker in expedient than intelligent in action. The first will have a taste for history, the second for romance.

¶ 357. Large thumb. With a large thumb, a woman will be sensible and cautious in affairs of the heart. Love is with her a "goodly estate" and not a passion. She will be more apt to make a *mariage de convenance* than if her thumb be small. She will be sagacious, easy of conquest, *or else* unapproachable. There is no medium, for she will never descend to coquetry or jealousy.

¶ 358. Small thumb.

Weak hand. If a woman have a small thumb, she will be more capricious, more *coquette*, more prone to jealousy, more fascinating, and more seductive than if she have a large thumb. If the rest of her hand is weak, her character may also be weak, uncertain, irritable, and careless; now enthusiastic, now despondent and apathetic; whilst, confiding and naïve, it is impossible for her to keep a secret. With her, love is a passion, an emotion, powerful and fervid. She will demand an undivided fidelity, and a sentimental, romantic form of adoration.

¶ 359. Absence of the elementary type The elementary hand is never [or hardly ever] found amongst women. Their natural intelligence, the cares of maternity, the exquisite and complicated physical constitutions of women requiring a higher instinct, a greater intuitive intelligence than is ever constitutionally necessary to man. Consequently, in countries where, amongst the men, the elementary type predominates, the women always have the upper hand, and direct the affairs of their husbands and other "concomitant males."

¶ 360. Feeble hands.
Women who live only an objectless, butterfly life of pleasure, love, and luxury, have small conic hands, soft and rather thick.

¶ 361. English hands.
English women, taking, as they do, so large a part in the administration and arrangement of household affairs, have their fingers, for the most part, delicately squared.

¶ 362. Female curiosity
Women who have that inherent curiosity which has been set down as the exclusive right of their sex [but which I have heard defined in self-defence by the more scientific phrase "excess of sympathy"] have their fingers so much of different forms and shapes, that, when the fingers are closed together and held up against the light, there are chinks and crannies between them through which the light is visible. When, on the other hand, the fingers fit so tightly against one another as to show no light between them when so held up, it is a sign of avarice and meanness, or, at any rate, of want of generosity. And these last two indications apply equally to men [*vide* ¶ 152].

¶ 363. Spatulate fingers and small thumb.
Women with spatulate fingers and a little thumb are warm friends, affectionate and impulsive, unreserved and active, fond of exercise, of animals, and of witnessing feats of skill or strength. Their needlework is useful and complete rather than artistic and showy, and they like to manage and make much of children, whether their own or other people's.

¶ 364. Square fingers and small thumb.
With square fingers and a small thumb we get punctuality, order, and arrangement in household affairs, a *ménage* well regulated and neatly appointed, and a highly developed instinct of real life and of the things which make it tolerable. Square-fingered women require courtesy, order, and regulation in

affairs of the heart; they like men to be distinguished without being eccentric, spirited without being wild, quiet, self-confident, and self-contained, untinged by jealousy or inconstancy; they are particularly careful of social observance, and fly from anything extraordinary, or worse still, *vulgar!* If, however, the squareness is too pronounced we find a fussy, disciplinarian disposition, eminently respectable, but horribly irritating to live with.

¶ 365. Large thumb. With a large thumb and square or spatulate fingers we find the tyrannical, "worrying" woman, impatient of control, loud-voiced, and abusive of power when it is entrusted to her.

¶ 366. "Pretty" hands. A little rosy, soft, smooth hand, thin, but not bony, and with little joints slightly developed, indicates a vivacious, sparkling little woman. To win her you must be bright, clever, witty, spontaneous, amusing; in affairs of *la grande passion* you must be quick and sparkling, rather than romantic and sentimental, as you must be with the conic-fingered woman.

¶ 367. Conic fingers. With the latter you must be ardent, timid, self-assured, humble; explaining, excusing, justifying all things. Such women are generally indolent, fantastic, and strongly inclined to sensuality.

¶ 368. Small pointed hands. With slight, smooth, pointed fingers, a small thumb, and a narrow palm; we find in the subject the highest romanticism and ideality as regards affairs of the world [for which they are eminently unsuited] and of the heart [in which their ideal is never attained]. Pleasure is with them more a matter of heart and soul than of physical emotion: they combine fervour and indolence, and they have the utmost disregard for the conventionalities and realities of life; they are more prone to excessive piety and superstitious worship than to real devotion. Genius is a thing with them infinitely superior to common sense, and from the height of their radiant idealism they look down upon all intelligences of the beautiful in the abstract which are less sublime than theirs.

Thus could I discourse through many sub-sections on the more subtle and interesting interpretations of the female hand; but time, space, and an appreciation of my readers' mental capacity deter me. With the data I have given above the student of cheirosophy will easily learn to distinguish between the indications of identical formations according as he finds them on the male or on the female hand.

¶ 869. Afterwords

AFTERWORDS.—I have therefore now, in the compass of three sub-sections, presented the student with all such outlines as are necessary for him to learn, of the important first section of the science of cheirosophy. A few moments' reflection will convince him of the vital necessity for mastering this branch of the science of the hand, and of combining it inextricably with the practice of the more profound cheiromancy, to which we are about to turn. I have throughout this section kept my notes as free as possible from intercalations of the modifying influences of the mounts and lines of the palm, but now that the student is well grounded in the indications of the outward forms of the hands I shall be able constantly to refer back to this preceding section, in search of modifications for the indications afforded by cheiromancy proper; and in this way the inseparability of the two sections will be more than ever made manifest to the student.

SECTION II.

Cheiromancy.

1. J. Hartlieb—" Die Kunst Ciromantia." (Augsburg, 1475.)
2. Aristotle—"Cyromantia Aristotelis cum Figuris." (Ulme, 1490.) [MS. Brit. Mus.]
3. B. Cocles—" Le Compendion et Brief Enseignement de Physiognomie et Chiromancie." (Paris, 1550.)
4. " La Chiromence de Patrice Tricasse des Ceresars," etc. (Paris, 1552.)
5. Tricassi Cerasarensis—. . . Enarratio pulcherrima principiorum Chyromantiæ." (Noribergæ, 1560.)
6. " Cheiromantiæ Theorica Practica Concordantia genethliaca Vetustis novitate addita." (Erphordiæ, 1559.)
7. J. Rothmannus—" Keiromantia, or the Art of Divining by the Lines, etc., engraven in the Hand of Man."— *Translated by G. Wharton.* (London, 1651.)
8. J. Indagine—" The Book of Palmestry and Physiognomy," etc.—*Englished by Fabian Withers.* (London, 1651.)
9. " Palmistry, the secrets thereof disclosed." (London, 1664.)
10. R. Saunders—"Physiognomie and Ciromancie." (London, 1671.)
11. A. Desbarrolles—" Les Mystères de la Main, revélés et expliqués." (Paris, 1859.) 16*th edition*, N. D.
" Révélations complètes : suite et fin." (Paris, 1874.)
12. R. A. Campbell—"Philosophic Chiromancy, Mysteries of the Hand revealed and explained." (St. Louis, 1879.)
13. H. Frith and E. Heron-Allen—" Chiromancy, or the Science of Palmistry," etc. (London, 1883.)
14. A. R. Craig—" Your luck's in your Hand ; or, the Science of Modern Palmistry " (London, N. D.); *Third edition of* " The Book of the Hand" (London, N. D.) ;
15. " Dick's Mysteries of the Hand" (New York, 1884.)

SECTION II.

CHEIROMANCY; OR, THE DEVELOPMENTS AND LINES OF THE PALM.

CHEIROMANCY

Νῦν δ' ἐπ' ἀδήλοις οἶσι τοῖς ἀπὸ τούτων ἐμαυτῷ γενησομένοις, ὅμως ἐπὶ τῷ συνοίσειν ὑμῖν, ἐὰν πράξητε, ταῦτα πεπεῖσθαι λέγειν αἱροῦμαι. Νικῴη δ' ὅ τι πᾶσιν ὑμῖν μέλλει συνοίσειν.

ΔΗΜΟΣΘΕΝΟΥΣ ΚΑΤΑ ΦΙΛΙΠΠΟΥ. Α'., 51.

¶ 370. Forewords.

FOREWORDS.—Cheiromancy is that branch of the science of cheirosophy which reveals not only the habits and temperaments of men, but also the events of their past, the conditions of their present, and the circumstances of their future, lives, by the inspection and interpretation of the formations of the palm of the hand, and the lines which are traced thereon. We have seen [¶¶ 106-8] how necessary it is that in making a cheirognomic examination of a subject the inspection should be conducted with a due regard to the cheiromancy of the hands; it will be seen immediately how much more important it is that the shapes of the hands and fingers should be considered in giving a cheiromantic explanation of any submitted palm. For what is clearer and more easily to be understood than that the character and temperament of a man [chiefly revealed by the cheirognomic examination of his hands] should very greatly influence, even if it does not absolutely bring about, the events which are recorded in his palms, so that, as I have said [¶¶ 106-8], a glance at the fingers and thumb will nearly always explain anything which appears doubtful in the palm, and by making a preliminary cheirognomic examination of a subject, the cheiromantic examination will be rendered very much clearer and easier of interpretation. Therefore, as we shall immediately see, I shall combine cheirognomy with cheiromancy far more than I combined cheiromancy with cheirognomy, [because we had not yet entered upon the consideration of this more complicated branch,] with a view to rendering my exposition easier of remembrance.

We shall consider in turn the mounts and the lines of the palm, with the signs and other modifications which it is necessary to bear in mind; but first, we must arrive at a complete understanding of the various parts of the hand, of the lines traced in the palm, and of the names by which they are known to cheirosophists.

¶ 371.
Astrologic names of the Mounts.
And I take this opportunity of pointing out that the names given to the Mounts [those of the principal planets] are not given to them by reason of any astrologic signification which they were at one time supposed to bear, but because we have been accustomed to connect certain characteristics with certain gods of the pagan mythology, and because it is therefore convenient to give to the formations of the hand which reveal certain characteristics, the names of the particular gods whose characteristics those were; a principle obviously more reasonable than to describe geographically in every instance the locality [in the hand] of the formation which it is desired to designate; a course which would inevitably culminate in a confusion only to be expected from the continual reitera-

tion of an indicative verbosity. I shall therefore be understood not to be using the expressions in the old astrologic sense when I make use of such terms as "The Mount of Venus," or "The Plain of Mars," but merely to be indicating the characteristic betrayed by a development of the hand at a certain point.

SUB-SECTION I.

AN EXPLANATION OF THE MAP OF THE HAND [PLATE VII.].

" Est pollex Veneris ; sed Juppiter indice gaudet,
Saturnus medium ; Sol medicumque tenet,
Hinc Stilbon minimum ; ferientem candida Luna
Possidet ; in cavea Mars sua castra tenet.[112]

¶ 372.
The map of the hand.
On Plate VII. you will find a complete map of the hand, whereon is written the specific and technical name given to each part thereof, the mounts being indicated in their proper position by the planetary signs [♀. ♃. ♄. ☉. ☿. ♂. and ☽.] for the sake of brevity and clearness.

¶ 373.
The thumb and Mount of Venus.
The thumb is consecrated to Venus [♀], and at its base will be found the Mount of Venus, surrounded

[112] These verses are of immense antiquity, and I quote them as a heading to this sub-section, as they convey all that is required to be known as to the localities of the mounts in the palm. Their translation is as follows :—" The thumb is the finger of Venus, but Jupiter delights in the first finger [*indice*], Saturn holds the second finger [*medium*], and Apollo the third [*medicum*]. Mercury [*Stilbon*] is here at the smallest finger, and the chaste Moon occupies the percussion [*ferientem*] ; in the hollow of the hand Mars holds his camp." Some of these terms require explanation : "*indice*," the first or index-finger, so called from its use in pointing or *indicating;* "medium," the middle finger, so called by persons who count the thumb as one of the *five*

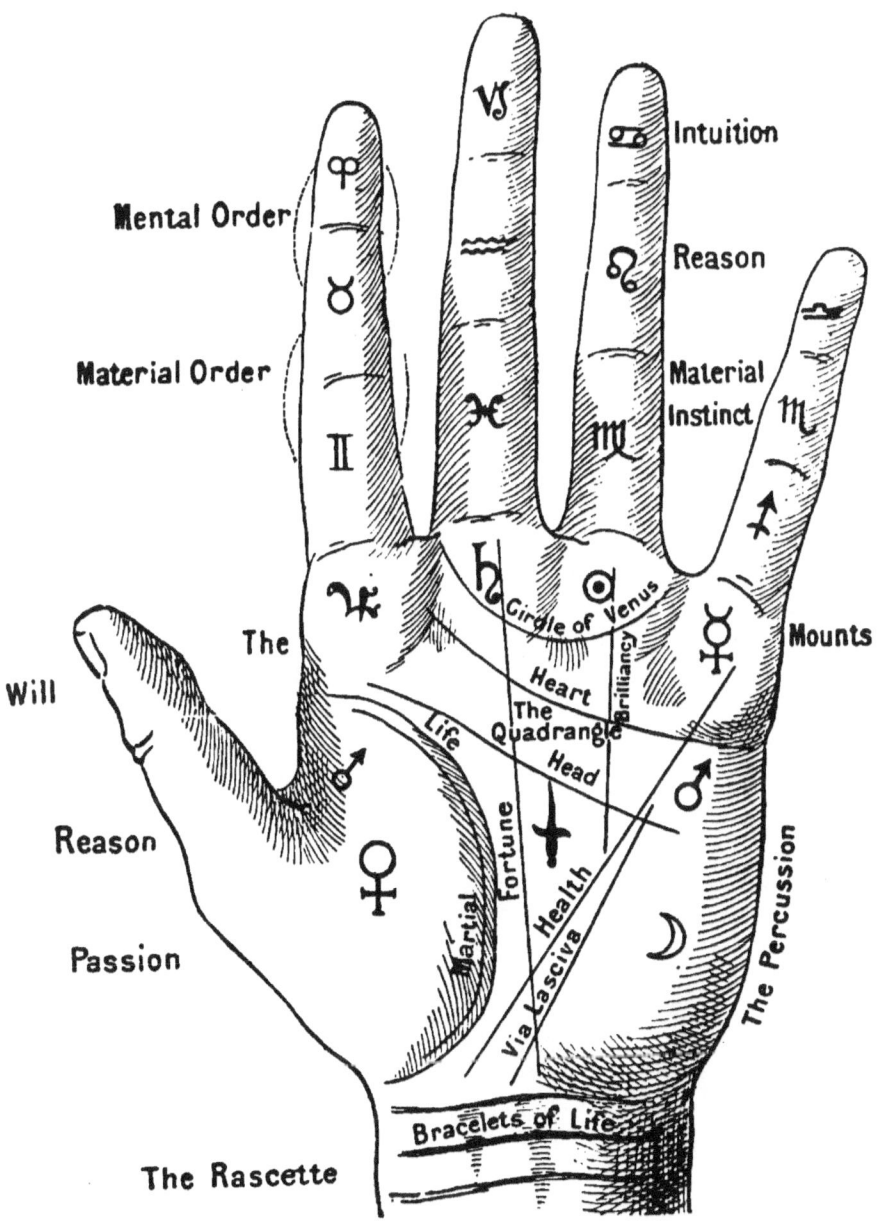

PLATE VII.—THE MAP OF THE HAND.

by the line of life. The base, or "ball" of the thumb, is frequently looked upon as a phalanx distinct from the hand, but, cheirosophically speaking, the thumb has but two phalanges [*vide* S. I. ss. II., § 5, ¶¶ 184-5], this base being termed the Mount of Venus.

The first finger [or index] is that of Jupiter [♃], and at its base [*i.e.*, immediately below it, at the top of the palm] will be found the Mount of Jupiter.

¶ 374. First finger Jupiter.

The second finger [or middle finger] is that of Saturn [♄], and the mount which should be found immediately below it is the Mount of Saturn.

¶ 375. Second finger Saturn.

The third finger [or ring finger] is termed the finger of Apollo [☉] [or of the Sun], and the Mount of Apollo will be found, if present, at its base.

¶ 376. Third finger Apollo.

The fourth finger [or little finger] is that of Mercury [☿], whose mount will in like manner be found immediately beneath it.

¶ 377. Fourth finger Mercury.

Just below the Mount of Mercury [between the line of heart and the line of head] is the Mount of Mars [♂].

¶ 378. Mount of Mars.

Underneath this last mount, and extending from it to the wrist, is found the Mount of the Moon [☽].

¶ 379. Mount of Moon.

The whole of the centre of the palm is occupied by the Plain or Triangle of Mars, which is comprised between the line of life, the line of head, and the Mounts of Mars and of the Moon.

¶ 380. Plain of Mars.

fingers; "medicum," the third finger, so called because it was always used by doctors in mixing their drugs, for reasons given at length in another place [*vide* p. 41, ¶ 37]. "Stilbon," the little finger, *i.e.*, that of Mercury. Stilbo was a name often given by the ancients to the planet Mercury, in consequence, Lemprière says, of its shining appearance [*vide* hereon Cicero, "De Naturâ Deorum," Book II., cap. 20 : "Infra hanc autem stella Mercurii est : ea Στίλβων appellatur a Græcis "] ; "ferientem," the percussion; the *outside* of the hand, which is termed in modern cheirosophy the *percussion*, and in ancient cheirosophy *feriens*, on account of its being the part usually struck on the table in argument, or generally used in driving anything into its place.

¶ 381. **The triangle.** This part of the hand is also called the triangle, and is composed of the upper angle—*i.e.*, that formed by the junction of the lines of life and of head; the inner angle—*i.e.*, that formed by the junction of the line of head with the line of health or the line of fate, at the Mount of the Moon; and the lower angle, which is formed by the approximation or junction of the line of life and the line of health [when the latter is present].

¶ 382. **The quadrangle.** The quadrangle is the rectangular space comprised between the lines of head and of heart, and is generally bounded on the one side by the line of fate and on the other by the line of Apollo.

¶ 383. **The rascette.** The rascette or restreinte is the point on the wrist at which it joins the hand, which is generally occupied by one or more lines, which are more or less apparent, the upper one of which is known as the rascette and the others as the restreintes, the whole forming what are called the Bracelets of Life.

¶ 384. **Line of life.** The lines generally found in the hands are as follows:—The line of life, which encircles the ball of the thumb or Mount of Venus;

¶ 385. **Line of head.** The line of head, which, starting from the beginning of the line of life [to which it is usually joined], between the thumb and first finger, runs straight across the hand;

¶ 386. **Line of heart.** The line of heart, which, starting from the Mounts of Jupiter or of Saturn, runs across the hand immediately below the Mounts of Saturn, Apollo, and Mercury, ending at the percussion;

¶ 387. **Line of fortune.** The line of fate or fortune, which, starting either from the line of life, from the rascette, or from the Mount of the Moon, runs up more or less directly to the middle finger [the finger of Saturn];

¶ 388. **Line of health.** The line of health or liver, which, starting near the wrist, at the base of the line of life, rises diagonally across the hand to meet the line of head, close to the

Mount of Mars, or at the top of the Mount of the Moon; *and*

The line of art and brilliancy, which, rising from the triangle or its vicinity rises to the finger of Apollo [the third], cutting across the mount at its base.

¶ 389. Line of Apollo.

To these are added three lesser lines sometimes found in a hand, which are:—The line of Mars, which lies close inside the line of life, which it follows as a sister line [*vide* ¶¶ 531 and 549];

¶ 390. Line of Mars.

The ring or girdle of Venus, which encloses the Mounts of Saturn and of Apollo; *and*

¶ 391. Girdle of Venus.

The Via Lasciva, or milky way, which, rising from the wrist, traverses the Mount of the Moon.

¶ 392. Via Lasciva.

The principal lines are also known by other technical names, which [to avoid repetition] will sometimes be used in the following pages. Thus the line of life is also called the Vital. The line of head is also called the Natural. The line of heart is also called the Mensal. The line of fortune is also called the Saturnian. The line of art or brilliancy is also called the Apollonian, and the line of health is often known as the Hepatic.

¶ 393. Equivalent names.

The ancient Cheiromants used also to consider the twelve phalanges of the fingers, as representing the twelve signs of the Zodiac, and used therefrom to predict the seasons at which certain events would come to pass. This is a branch of cheirosophy which, it is needless to say, is now obsolete, having been refined away with the rest of the dross which used to disguise the pure metal of the science; but Miss Horsley has put them into the diagram at my request, as they may be interesting to my readers.

¶ 394. The signs of the Zodiac.

Having, therefore, mastered what may be called the geography of the hand, we can now turn to the consideration of the cheiromancy of the hand, commencing, as I have said, with the mounts, and continuing with the lines; but before entering into the

¶ 395. The study of cheiromancy.

minute discussion and examination of each particular mount and of each particular line, I wish to devote a sub-section to the enunciation of certain general principles, which, applying to all mounts and lines equally, must be carefully borne in mind throughout every cheiromantic examination.

¶ 396.
Modus operandi

Note.—It has been suggested to me that I should have opened this work with a sub-section, devoted to an explanation of the *modus operandi* of cheirosophy, of the method in which the cheirosophist should proceed when he undertakes the cheirosophic examination of a subject; but I have reflected that it is of no use deferring the real business of the book by teaching my readers how to put into practice principles which they have not yet acquired—a thing paradoxical in itself. I have, therefore, relegated this matter to the conclusion of the work, as being the more appropriate position for such a sub-section.

SUB-SECTION II.

GENERAL PRINCIPLES TO BE BORNE IN MIND.

GENERAL PRINCIPLES.

§ 1. *As to the Mounts.*

THE MOUNTS

THE mount, which is the highest in the hand, will [as we shall see] give the keynote to the character of the subject, and will be the first thing sought for; and when the characteristics are thus pronounced by the development of a particular mount, the lesser [but still noticeable] development of another mount will indicate that the characteristics of the lesser will influence those of the greater, modifying, and in a manner perfecting, those of the reigning development.

¶ 397. The leading mount and the lesser mounts

You will seldom find that a subject has only one mount developed, and you must bear in mind in all cases that the modifying characteristics must be considered in reading the primary indications of the principal mount.

¶ 398. Modifying indications.

A characteristic betrayed by a prevailing mount can never lie dormant in a subject; opportunities

¶ 399. Action of a quality.

for exercising the qualities indicated will always arise, for the subject will, in a way, make their himself—*e.g.*, a man whose leading mount is that of Mars will, by provoking others, call the talent of his character into play.

¶ 400.
Equality of the mounts.

If a subject have *no* particularly prominent mount in his hand—*i.e.*, all the mounts are equal—you will find a singular regularity of mind and harmony of existence to be his lot.

¶ 401.
Absence of any mount at all.

If all the mounts are null, and the places where they should be are merely occupied by a plain or a hollow, you will find that the subject has never had any opportunity of developing any particular characteristic, and the life will be a purely vegetative one.

¶ 402.
Other indications of the leading mount.

A mount may, instead of being high, be *broad* and full, or it may be covered with little lines. These conditions of the mount give it the same effect as if it were highly developed; and it must be remarked that, if a mount is much covered by lines, it will

Excess.

betray an excess and over abundance of the qualities of the mount, which prove an insurmountable obstacle to the good effects thereof. Excess of a mount does not give *force*, but *fever* to its quality, producing monomanias, especially if the thumb and the line of head are weak.

¶ 403.
Lines on a mount.

One line upon a mount just emphasizes it enough to be a fortunate sign upon it; *two* lines show uncertainty in the operation of the qualities, especially if they are crossed; and *three*, except in some rare cases, give misfortune arising from the qualities of the mount, *unless* they be even, straight, and parallel. If no other mount is developed, the one upon which most lines are found will be the leading mount in the hand.

¶ 404.
Cross lines on a mount.

Lines placed *crosswise* upon a mount always denote obstacles, and seriously interfere with the goodness of

AS TO THE MOUNTS.

other main lines, which end upon the mount, as in the cases of the mounts and lines of Saturn, or of Apollo [*vide* pp. 206 and 211], *unless* the ascending line is deeper than the cross lines, in which case the evil indications of the cross lines are destroyed.

¶ 405. Capillary cross lines.

De Peruchio [*vide* Note [98], p. 116] affirms that little capillary cross lines upon a mount signify wounds; thus on the Mount of Jupiter they signify a wound to the head; on that of Saturn, to the breast; on that of Apollo, to the arms; on that of Mercury, to the legs; on that of Venus, to the body. I have encountered some strange confirmations of this statement, but such instances are rare.

¶ 406. Modifications by lines.

Thus it will be seen that the indications afforded by any particular mount may be greatly modified, if not annulled, by the appearance of lines upon it, or in its immediate vicinity, so that these must be carefully sought for and examined concomitantly.

¶ 407. Displacement of the mounts.

It will be very frequently found that the mounts are not exactly under the fingers, but lean, as it were, in the direction of the neighbouring mount. In such cases the prevailing development takes a modification from that towards which it inclines.

¶ 408. Good or bad influences of a mount.

Finally, the influence of the mount, which is principally developed, may be either good or bad. Desbarrolles has stated that this may be determined by the expression of the face, but I think by far the surer and more scientific determination may be arrived at by inspecting the formation of the tips of the fingers, the consistency of the hand, and the development of the thumb. Thus, pointed fingers reveal an intuition, a lofty *idealism* of the quality. Square fingers will look at the *reasonable* aspects of character, and spatulate will cultivate the *material* qualities of the mount—*e.g.*, Jupiter developed will indicate, with pointed fingers, religion; with square fingers, pride; and with spatulate fingers, tyranny.

Apollo developed will indicate, with pointed fingers, love of glory; with square fingers, realism in art; and with spatulate fingers, love of wealth and luxury. And so on with the other mounts.

¶ 409. *Phrenology and physiognomy.* Most authors have gone into the phrenology and physiognomy characteristic of each type, but as I consider this to be not only confusing, but irrelevant to the study of pure cheiromancy, I have avoided the consideration of this matter.

THE LINES.

§ 2. *As to the Lines.*

¶ 410. *Proper appearance.* The lines in a hand should be clear and apparent. They should be neat and well coloured [not broad and pale], free from branches, breaks, inequalities, or modifications of any sort, except in some few cases, which will be pointed out in due course. A broad pale line always signifies [by indicating excess] a defect of, and obstacle to, the natural indications and qualities of the line.

¶ 411. *Pale lines.* Pale lines signify a phlegmatic or lymphatic temperament, with a strong tendency towards effeminacy [women nearly always have very pale lines]. Such subjects are easily put out, and as easily calmed again; they are generally liberal, and subject to strong enthusiasms, which are of short duration.

¶ 412. *Red lines.* Red lines indicate a sanguine temperament, and are good; such subjects are gay, pleasant in manner, and honest.

¶ 413. *Yellow lines.* Yellow lines denote biliousness and feebleness of the liver; such subjects are quick-tempered, prompt in action, generally ambitious, vigilant, vindictive, and proud.

¶ 414. *Livid lines.* Livid lines, with a tendency towards blackishness, betray a melancholy and often a revengeful disposition. Such subjects are grave in demeanour and cunning in character, affable, but haughty; and these indications

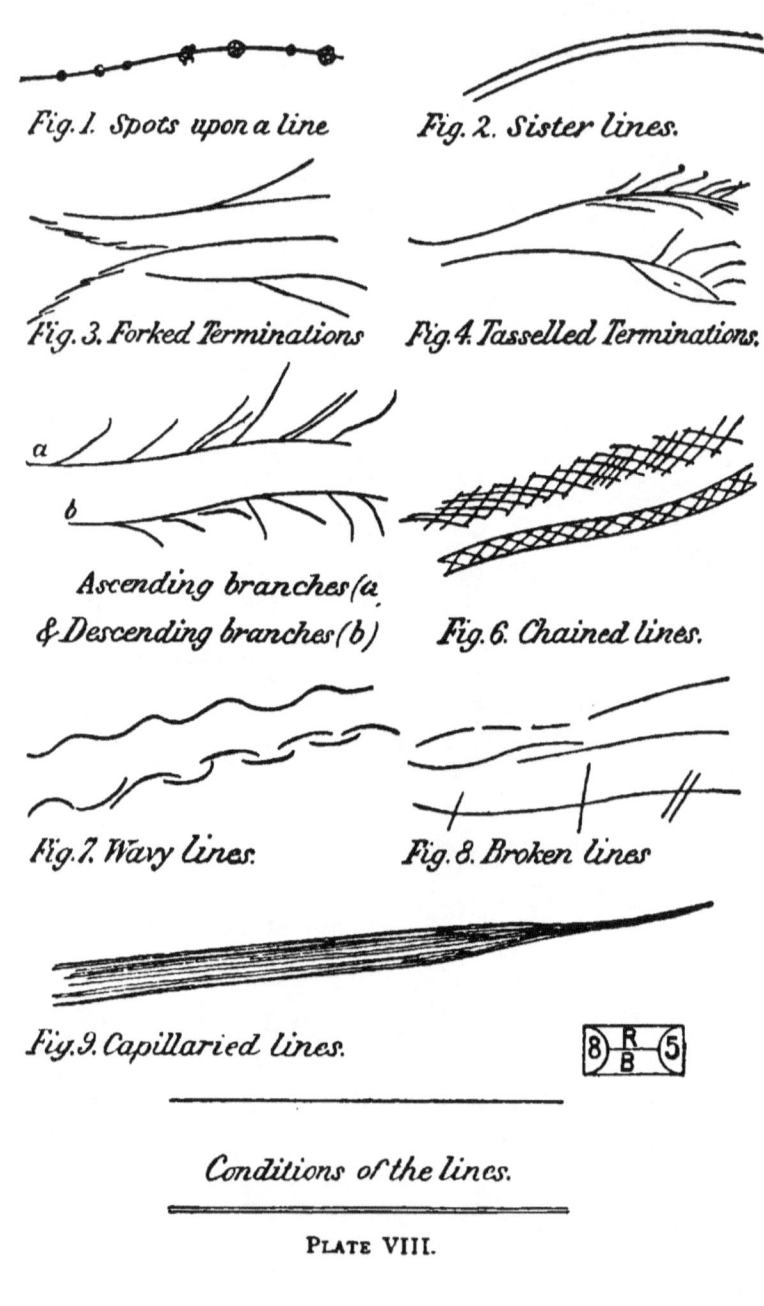

Fig. 1. Spots upon a line   Fig. 2. Sister lines.

Fig. 3. Forked Terminations   Fig. 4. Tasselled Terminations.

Ascending branches (a) & Descending branches (b)   Fig. 6. Chained lines.

Fig. 7. Wavy lines.   Fig. 8. Broken lines

Fig. 9. Capillaried lines.

Conditions of the lines.

PLATE VIII.

are the more certain if the fingers are long and the thumb is broad.

¶ 415. Black spots on a line.

Black spots upon a line indicate always nervous diseases, whilst livid holes betray the presence of an organic affection of the part corresponding with the line [Fig. 1, Plate VIII.].

¶ 416. Feeble mounts.

It must be noted that, however well coloured lines may be, a feeble development of the mounts will counteract their good indications.

¶ 417. Persons born by night or by day

The ancient cheiromants used to affirm that people who had been born in the daytime had the lines clearer marked in the right hand, whilst those who had been born in the night had them more apparent in the left; a statement which we must class with the dicta laid down for ascertaining birthdays noted in another place [*vide* Note [98], p. 116]; they also stated that the former resembled their fathers, whilst the latter took after their mothers.

¶ 418. Alteration of lines.

It must also be noted that lines may enlarge, diminish, and even disappear, so that the province of the cheirosophist is, as I have said below, [*vide* ¶ 814,] to indicate the present condition and indications of the lines, and the likelihood of their future modification. There is one thing to be noted in connection with this matter, which is, that the indications of cunning never alter or become modified; cunning being a characteristic which is acquired, and a characteristic thus acquired is never lost by a weak character on account of inability to free itself, nor by a strong one from a disinclination to do so.

¶ 419. Necessity for appearance of indications in *both* hands.

Again, in reading the lines a *single* indication must never be accepted as final, especially if it is a bad one. To make *any* indication certain [whether good or bad] corroborating signs must be sought for in both hands, and the absence of corroboration in one hand will contradict, or at any rate greatly modify, any evil sign in the other. A single sign only affords a presumption of the tendency or event which it indicates, and the cause of the danger must be found in the aspect of the mounts, and other lines of the palm, or the development and formations of the whole hand. In the same way the indication of prudence in the second joint of the thumb will go far towards modifying an evil prognostic, which may be found in the palm.

¶ 420. Sister lines. When any principal line is accompanied throughout its course by a second line lying close to it, the principal line is greatly strengthened and benefited by this "sister line," as it is called. The consecutiveness of the sister will contradict the evils foreshadowed by a break in the principal line, [vide ¶ 531], but if *both* are broken, the evils are the more certainly to be feared [Fig. 2, Plate VIII.].

¶ 421. Very many lines in the palm. If the hand is covered with a multiplicity, a network of little lines which cross one another in all directions, it betrays a mental agitation and dissatisfaction with one's surroundings and oneself. It is always the outcome of a highly nervous temperament; and in a soft spatulate hand these little lines denote hypochondria.

¶ 422. Fork at the end of a line. A fork at the end of a line is often a good sign, for it increases the powers of the lines without carrying them too far. At the same time it often indicates a duplicity in connection with the qualities of the line [vide ¶ 589] [Fig. 3].

¶ 423. Tasselled at end. When the fork is reduplicated so as to form a tassel at the end of the line, the indication is bad, denoting feebleness and nervous palpitation of the organ represented [Fig. 4].

¶ 424. Ascending and descending branches. All branches *rising* from a line increase its good indications, whereas all *descending* branches accentuate its bad qualities. Ascending branches indicate richness, abundance of the qualities appertaining to a line; thus on the line of heart they denote warmth of

Fig. 10. The Star

Fig. 11. The Square

Fig. 12. The Spot

Fig. 13. The Circle

Fig. 14. The Island.

Fig. 15. The Triangle

Fig. 16. The Cross

Fig. 17. The Grille

Signs found in the Hand.
PLATE IX.

PLATE IX.

AS TO THE LINES.

affection and devotion; on the line of head they denote cleverness and intelligence; on the line of Saturn they denote good luck, and so on. These branches, when present, are nearly always found at the beginnings and endings of lines [Fig. 5].

A chained formation of a line indicates obstacles, struggles, and contrarieties of the characteristics afforded by it [Fig. 6].

¶ 425.
Chained lines

A wavy formation [Fig. 7] of a line signifies ill-luck, as does also a break in it. Breaks may be either simple interruptions or cessations of the line, or bars across it: they are always a bad sign, and the interrupting influence must be carefully sought [Fig. 8].

¶ 426.
Wavy lines.

When a line, instead of being single and clear, is composed of a number of little capillaries, which here and there, or at the ends, unite to form a single line, it betrays obstacles and ill-success, in the same way as chained lines [Plate VIII., Fig. 9].

¶ 427.
Capillaries.

I have gone into these general principles at this point, because it can never be too early to point them out. The reader will understand them better, however, if he will return to them after a perusal of the two succeeding sub-sections.

SUB-SECTION III.

THE MOUNTS OF THE HANDS.

THE MOUNTS.

¶ 428.
Their positions.

I HAVE indicated at ¶¶ 373-9 where the various mounts should be placed. The prevailing mount is the first thing to be observed in the palm of a hand, and it must be sought for with a careful regard to the general principles laid down in Sub-section I., § 1. In this sub-section we shall carefully consider the indications afforded by each mount in succession, as well as those of some of the principal combinations of mounts.

MOUNT OF JUPITER.

¶ 429.
Indications of the Mount.

§ 1. *The Mount of Jupiter* [♃].

The predominance of this mount in a hand denotes a genuine and reverential feeling of religion, a worthy and high ambition, honour, gaiety, and a love of nature. It also denotes a love of display, of ceremony, and of pomp, and is, consequently, generally developed in the hands of public entertainers of any sort. Such subjects talk loudly, are extremely self-confident, are just and well-minded, gallant and extravagant, and are always impetuous without being revengeful. These subjects are fond of flattery and fond of good living. They generally marry early, and are always well-built and handsome, having a certain *hauteur*, which enhances their charms without detracting from their good nature.

An excessive development of the mount will give arrogance, tyranny, ostentation, and, with pointed fingers, superstition. Such subjects will be votaries of pleasure, and vindictive, sparing nothing to attain their selfish ends.

¶ 430. Excess of the mount

If the mount is absent [*i.e.*, replaced by a cavity] the subject is prone to idleness and egoism, irreligious feelings, want of dignity, and a license which degenerates into vulgarity.

¶ 431. Absence of the mount.

The development of this mount gives to square fingers a great love of regularity and established authority. To long smooth fingers it imparts a love of luxury, especially if the fingers are large at the third phalanx [*vide* ¶ 146]. This mount *ought* always to be accompanied by a smooth, elastic, firm hand [not too hard], with a well-developed first phalanx to the thumb [Will].

¶ 432. With square and smooth fingers.

If to the good indications of this mount a favourably developed finger, or Mount of Saturn, be added, the success in life and good fortune of the subject is certain; Saturn denoting fatality, whether for good or evil.

¶ 433. Influence of Saturn.

A single line upon the mount indicates success. Many and confused lines upon the mount betray a constant, unsuccessful struggle for greatness, and if these confused lines are crossed, they denote unchastity, no matter the sex of the subject.

¶ 434. Lines on the mount.

A cross upon the mount denotes a happy marriage, and if a star be found there as well as the cross, it indicates a brilliant and advantageous alliance.

¶ 435. Cross and star on the mount.

A spot upon the mount indicates a fall of position, and loss of honour or credit.

¶ 436. Spot.

¶ 437. *Its religion.* A long thumb and a development of the first joint in the fingers will give to this mount free thought and irreverence in religion. If, besides these, we find pointed fingers and what is called the "Croix Mystique," you will find ecstasy in matters religious, tending even to fanaticism.

¶ 438. *Displacement of the mount.* If, instead of being in position immediately underneath the finger of Jupiter, [or forefinger,] the mount is displaced and inclines towards that of Saturn, it acquires a serious tone and demeanour, and gives a desire for success in science, theology, or classical scholarship.

¶ 439. *Combinations of the mount with others.* If with the Mount of Jupiter we find also the Mount of Apollo [☉] developed, it indicates good fortune and wealth. Combined with the Mount of Mercury [☿] we find a love of exact science and philosophy. Such subjects are inclined to be poetic, are well behaved and clever; they make the most successful doctors. To a bad hand this combination will give vanity, egoism, a love of chatter, fanaticism, charlatanry, and immorality. *Mars.* Combined with the Mount of Mars [♂] it gives audacity and the talent of strategy. Such subjects are self-confident, successful, and fond of celebrity. To a bad hand such a combination gives insolence, ferocity, revolt, dissipation, and inconstancy. *Moon.* A combination of the Mounts of Jupiter and of the Moon [☽] makes a subject honourable, placid, and just. *Venus.* With the Mount of Venus [♀] a subject of this Jupiterian type becomes sociable, simple-minded, gay, sincere, fond of pleasure, and generous. If the hand is, on the whole, bad, the combination will denote effeminacy, feeble-mindedness, caprice, and a love of debauch.

MOUNT OF SATURN.

§ 2. *The Mount of Saturn* [♄].

¶ 440. *Effects of the mount.* The predominance of this mount in a hand denotes a character in which to prudence and natural caution

THE MOUNT OF SATURN.

is added a *fatality* for good or evil, which is extreme.[113] Such subjects are always sensitive and particular about little things, even though their fingers be short [vide ¶¶ 131-2]. The mount also denotes a tendency to occult science, to incredulity, and to epicureanism of temperament. Such subjects are always inclined to be morbid and melancholy. They are timid, and love solitude, and a quiet life in which there is neither great good fortune nor great ill fortune; they are also fonder of serious music than of gay melody. They take naturally to such pursuits as agriculture, horticulture, or mineralogy, having a natural *penchant* for anything connected with the earth. These subjects seldom marry, are extremely self-centred and self-confident, and care nothing for what other people may think of them.

¶ 441. Excessive development of the mount.

The mount is seldom *very* high, for fatality is always, to a certain extent, modifiable; but when there is an excess of formation on this mount it betrays taciturnity, sadness, an increased morbidity and love of solitude, remorse, and asceticism, with the horrible opposing characteristics of an intense fear and horror of death, with a morbid tendency to, and curiosity concerning suicide. The evil indications of an excessive

---

[113] By fatality is meant *certainty, i.e.*, the indications of the middle finger are always looked upon as certain and unavoidable; and we find a curious parallel to this in comparative anatomy by observing the greater *constancy* of this finger: *i.e.*, the fact that it is the one digit that is never discarded in the brute creation. Professor Owen calls attention to this, in his work "On the Nature of Limbs" [p. 38], where he points out the fact that the extreme digits [the thumb and little finger] are the least constant and universal, whilst the "digitus medius" is the most constant of all in the vertebrate series, and most entitled to be viewed as the persistent representative of the *fingers;* it shows a superiority of size, though few would be led thereby to suspect that the bones forming the three joints of this finger answer to those of the foot of the horse, and that the *nail* of this finger represents the *hoof* in the horse, etc. [vide ¶ 33].

development may be greatly modified by a well-formed Mount of Venus [♀].

**¶ 442.
"Saturnian" hand.**
The Saturnian hand has generally long, bony fingers, which give it philosophy, the second finger [that of Saturn] is large, with the first [or nailed] phalanx highly developed, the mount, if not high, being generally strongly lined. A bad Saturnian hand has a hard rough skin and a thick wrist.

**¶ 443.
Absence of the mount.**
If the mount is quite absent the indication is of an insignificant, "vegetable" existence, unmoved by any great depth of feeling, and one which is continually oppressed by a sense of misfortune. But when it is thus absent it may be replaced by a well-traced line of fate [or Saturn].

**¶ 444.
Lines on the mount.**
A single straight line upon the mount signifies good fortune and success, whilst a plurality of lines thereon indicates a proportionate ill-luck. A succession of little lines placed ladder-wise across the mount and extending upon that of Jupiter indicate an easy and gradual progression to high honour.

**¶ 445.
Spot on the mount.**
A spot upon the mount always indicates an evil fatality, the cause of which must be sought for upon the lines of head or of fate.[114]

**¶ 446.
Branch from the line of heart.**
If a branch [not the end of the line of heart or of Saturn] rises from the line of heart on to the Mount of Saturn, it denotes worry, travail, and anxiety; if the branch is clean and single however, it will foreshadow wealth as a result of those anxieties [*l*, in Plate X.].

**¶ 447.
Displacement of the mount.**
If, instead of being in its proper position beneath the second finger, the mount is displaced towards Jupiter, it has the same significance as the displacement of the Mount of Jupiter towards Saturn [*vide* ¶ 438]. If, on the other hand, it is displaced towards the Mount of Apollo, it betokens a fatality which can be, and must be, striven against.

[114] Signum crucis in Monte Saturnio sterilitatem indicat.

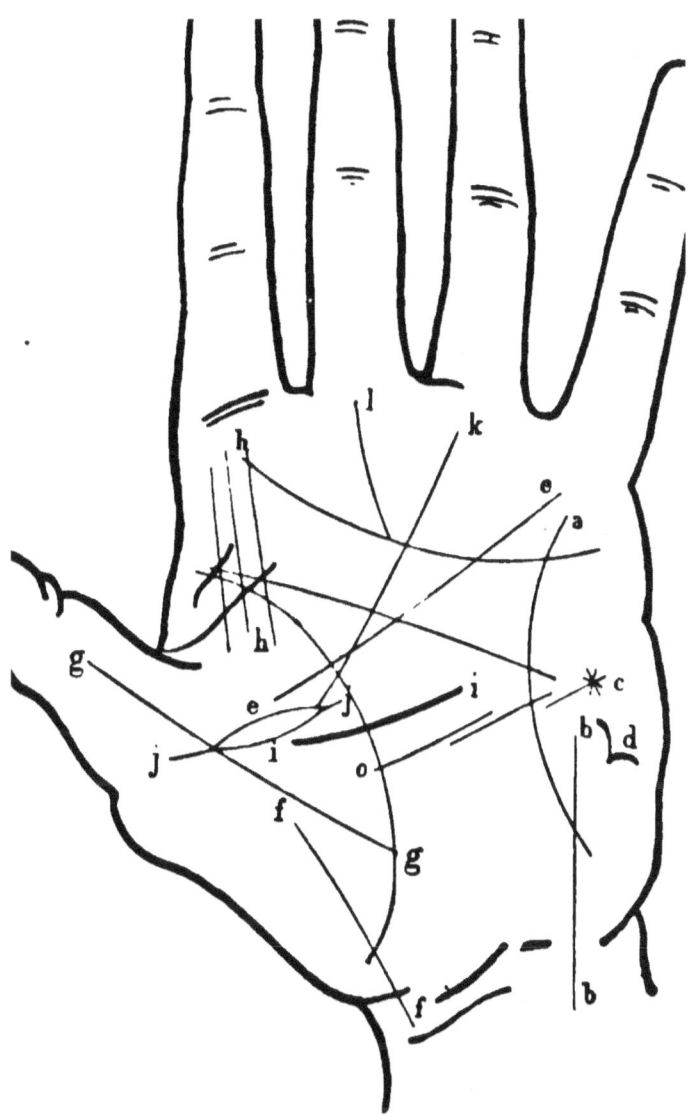

PLATE X.—LINES UPON THE MOUNTS OF THE PALM.

If, together with the Mount of Saturn, we find the Mount of Jupiter developed, we shall find gentleness, patience, and respect in a good hand, or want of appreciation, inability to make use of opportunities, melancholy, hysteria, and want of taste in a bad hand. Combined with that of Mercury this mount gives us antiquarian research, and love of science from an "amateur" point of view, a talent for medicine, and a desire for information on various subjects. Such subjects are clever at individualizing and classing, and are generally happy. And this latter indication generally holds good even when the rest of the hand is bad, in which case the combination of Saturn and Mercury gives us perfidy, perjury, sullen temper, revenge, theft, want of filial affection, and charlatanry.

¶ 448. Combinatio with other mounts. Saturn.

Mercury

With the Mount of Mars equally developed this mount betokens aggressiveness, bitterness of humour, a false superiority, insolence, immodesty, and cynicism. The combination of the Mounts of Venus and Saturn will give us a love of and a search after truth in matters occult, piety, charity, logic, self-control, with a tendency to jealousy and love of display. If the hand is bad the combination will betray frivolity, curiosity, and, if the Mount of Saturn be the more strongly developed of the two, we shall find pride, envy, and debauchery. When the Mounts of the Moon and of Saturn find themselves equally developed in a hand, we have a subject whose intuition and pure talent for occultism is remarkably developed. It is a curious fact that these latter subjects are generally frightfully ugly.

¶ 449. Further combinations Mars.

Venus.

Moon.

§ 3. *The Mount of Apollo* [☉].

MOUNT OF APOLLO.

A hand in which this mount is developed is essentially that of a subject whose prevailing tastes and instincts are artistic, and it always gives to its

¶ 450. Fortune of the mount.

possessor a greater or a less degree of success, glory, celebrity, and brilliancy of fortune, denoting, as it does, genius, intelligence, tolerance, and wealth, the characteristics of the type being self-confidence, beauty, grace, and tolerance in all things.

¶ 451. *Indications of the mount.* Such subjects are inventive and imitative, being often great discoverers. Their principal failings are, *quick* temper [though not of long duration] and a certain incapacity for very *close* friendships, though they are generally benevolent and generous, even devoted were it not for the inseparable strain of fickleness. Proud and eloquent on matters of art, they love anything which is brilliant, such as jewellery and the more ornamental forms of worship, for they are religious from a gratitude for blessings received rather than from a superstitious reverence. They make stern and unrelenting judges, and their love is more affectionate than sensual.

¶ 452. *Further indications.* These Apollonian subjects love to shine before the world, and not to be the cynosures of a small circle of admirers, though they hate the idea of ostentation or undeserved glory; they will not explain themselves in dogmatizing unless they think their audiences are sympathetic, refusing to waste words on ignorant cavillers, or to persuade people to accept their opinions. In marriage they are, unfortunately, very often unlucky, for their ideal, their standard of excellence, is unreasonably high.

¶ 453. *"Apollonian" hand.* The normal development of a hand bearing this mount high shows smooth fingers, with the tips mixed [*vide* p. 170] or slightly squared; the palm of an equal length with the fingers, a well-marked phalanx of logic, and either one very deep, or three strong lines upon the mount.

¶ 454. *Excess of the mount.* If the mount is developed to excess it indicates a love of wealth and of extravagance in expenditure, instincts of luxury, fatuity, envy, and curiosity, a quick, unreasoning temper, and a strong tendency to levity, frivolity, and sophistry. Such subjects are boastful, vain, think themselves unappreciated, but highly superior to their fellow-men; these are the poets, painters, and musicians and actors who dwell upon the loss suffered by the world in their *non*-appearance at St. James's Hall, Burlington House, and the Lyceum. This excessive development is generally accompanied, and is emphasized by, twisted fingers, spatulated soft hands, a grille [*vide* ¶ 701] on the mount, with a long phalanx of will and proportionately short phalanx of logic.

If, on the other hand, this mount is absent in both hands, its absence betrays materiality and indifference to matters artistic, giving a dull, unenlightened life.

¶ 455. Absence of the mount.

A single line deeply traced upon the mount indicates fortune and glory; two lines indicate considerable talent, but a great probability of failure, whilst many confused lines show a tendency to lean to the scientific aspects of art.

¶ 456. Lines of the mount.

If the mount is merely *developed*, having no line marked upon it, it shows a love of the beautiful, but not necessarily a talent for production of works of art.

¶ 457. Development.

A spot upon the mount denotes a grave danger of a loss of reputation or caste.

¶ 458. Spot on the mount.

When in a hand the Mounts of Apollo and of Mercury are found equally developed, we find a character in which justice, firmness, perspicacity, love of scientific research, combined with clearness of diction and eloquence, are salient features. The combination of Apollo and the Moon gives good sense, imagination, reflection, and light-heartedness. With an equal development of the Mount of Venus, we get amiability and a great desire to please.

¶ 459. Combination with other mounts. Mercury.

Moon.

Venus.

§ 4. The Mount of Mercury [☿].

MOUNT OF MERCURY.

¶ 460. Indications of the mount.

The pre-eminence in a hand of this mount indicates science, intelligence, spirit, eloquence, a capacity for commerce, speculation, industry, and invention, agility, promptitude in thought and action, and a *penchant* for travel and occult science.

¶ 461. Eloquence.

[The eloquence, which is one of the prevailing characteristics of the type, is of a kind denoted by the formations of the fingers. A high Mount of Mercury will give, with pointed fingers, brilliant oratory; with square fingers, clearness and reason in expounding; with spatulate fingers, force and vehemence in argument and dogma; with long fingers, details and parentheses; and with short fingers, brevity and conciseness. The great difference between the eloquence of these subjects, and of those whose prevailing mount is that of Apollo, is that the oratory of the former is sophistical and clever, rather than naïve and direct like that of the latter; it is this that makes them such good barristers. To assist their faculties in this respect still further, these subjects should always have short nails (*vide* ¶ 251)].

¶ 462. Indications of the mount.

Such subjects are good athletes, are agile, clever at games of skill, spontaneous in expedient, sharp in practice, with a great capacity for serious studies. Combined with these qualities we generally recognise envy, but amiability therewith; often [the other conditions of the hand being favourable] we find that these subjects are clever clairvoyants, seldom sensual, and generally good-humoured, and fond of playing with children so long as they are not otherwise seriously employed. This tendency to envy, by raising envious feelings at the aptitudes and successes of others, constantly drives these Mercurial subjects to take up and try a great variety of pursuits.

THE MOUNT OF MERCURY. 215

These subjects are great matchmakers, and frequently marry very young, choosing equally young persons for their helpmates.

¶ 463.
Marriage.

The normal development of the hand which accompanies this mount is as follows :—Long, smooth fingers, hard, slightly spatulated, [athletics,] or very soft with mixed tips [thought]. The finger of Mercury long and sometimes pointed. The high mount cut by a deep line, and the philosophic joint developed.

¶ 464.
Normal development of the type.

If the mount is developed to excess in a hand, it denotes theft, cunning, deceit, treachery, with pretentious ignorance. Such subjects are charlatans, running after the false and dishonest forms of occultism, and are generally superstitious. These hands usually have long twisted fingers, more or less turned back; soft hands, confused markings on the mount, and the phalanx of will long.[115]

¶ 465.
Excess.

A complete absence of the mount denotes inaptitude for science or for commercial enterprise.

¶ 466.
Absence.

A single line upon the mount indicates modesty and moderation, and in many instances a strange unexpected stroke of good fortune. A cross line extending upon the Mount of Apollo betrays charlatanry in science, and, in fact, the dishonest occultism I have alluded to above [¶ 465]. If this line have an "island" [*vide* ¶ 678] in it and cut the Line of Apollo [p. 261] or brilliancy, it denotes ill-luck, probably resulting from some perfectly innocent act.

¶ 467.
Lines on the mount.

Many mixed lines upon the mount denote astuteness and aptitude for sciences. If they reach as low as the line of heart, they denote liberality; and if to numerous rays on this mount a subject joins a high Mount of the Moon, his *penchant* for medical science will take the form of hypochondria. The elder

¶ 468.
Rayed mount.

---

[115] Mons Mercurialis prominens in manu qua Orbis Lacteus vel Via Lasciva apparet, lasciviam et voluptatis sordidum appetitum ostendit.

cheiromants have affirmed that a woman having this mount well rayed is sure to marry a doctor, or, at any rate, a man of science. If the lines on the mount merely take the form of little flecks and dashes, it is a practically sure indication of a babbling, chattering disposition.

¶ 469.
Lines on the side of the hand.
Lines on the percussion—*i.e.*, on the edge of the hand, between the base of the little finger and the line of heart—indicate *liaisons*, or serious affairs of the heart if horizontal, [*i.e.*, parallel with the line of the heart,] each line denoting a separate *liaison* or love affair, a single deep line denoting one strong and lasting affection."[116] If vertical they denote, almost invariably, the number of children which the subject has had. De Peruchio lays down the rule that if they are strong they denote boys, if faint, girls; and if they are short or indistinct the children are either dead or not yet born. Several vertical lines on the percussion, crossed by a line which starts from a star upon the mount, betrays sterility, whilst a marriage line, ending abruptly by a star, indicates a marriage or *liaison* of short duration, terminated by death.

¶ 470.
Smooth mount.
Grille.
Circle.
Spot.
The mount quite smooth and unlined indicates a cool, determined, and constant condition of mind. A grille upon the mount is a dangerous prognostic of a violent death, a circle also placed upon the mount indicating that it will be by water. A spot upon the mount indicates an error or misfortune in business.

¶ 471.
Effect of Apollo.
If the mount is high, and the hand contain a long line of Apollo, the commercial instinct will work itself out in speculation rather than in recognized and persevering commerce.

[116] Insula in qualibet brevium linearum quæ percussionem transcurrunt et stupii consuetudinem denotant [*vide* ¶¶ 678-9] ostendit consuetudinem cum propinquo vel etiam incestum. Insula in Lineâ Saturniâ denotat adulterium quod felicem vel infelicem fert fortunam sicut fausta vel infausta linea.

The mount leaning, as it were, towards that of Apollo, is a good sign, good enough to counteract a bad line of Saturn, [*vide* p. 256,] betokening science and eloquence. Leaning in a contrary direction [*i.e.*, towards the percussion,] it indicates commerce and industry.

¶ 472. Displacement of the mount.

Connected with the Mount of Venus by a good line, [*e e*, in Plate X.] this mount gives happiness and good fortune.

¶ 473. Connected with ♀.

Combined [*i.e.*, equally developed] with the Mount of Venus, we find wit, humour, gaiety, love of beauty, often piety, easy and sympathetic eloquence. In a bad hand [*i.e.*, if the fingers are twisted, the line of head weak, and the phalanx of will small] this combination will give inconsequence, contradiction, meddlesomeness, inconstancy, and want of perseverance. The combination of Mercury and Saturn in a hand is always good, giving to the sobriety and fatality of Saturn, a certain intuitive practicality which seldom fails to give good results. The Mount of Mercury is, however, one which is not often combined with the other mounts of the hand.

¶ 474. Combination with other mounts. Venus.

Saturn.

§ 5. *The Mount of Mars* [♂].

Mount of Mars.

The discussion of the Mount of Mars is not fraught with that simplicity which characterises that of the other mounts. It is, in a manner, divided into the Mount of Mars properly so called, which is situated, as may be seen, beneath the Mount of Mercury on the percussion of the hand; and that development or extension of the mount into the palm of the hand [shown in Plate VII. by a dagger] which is known as the Plain of Mars. It will be seen that a development of the *Mount* of Mars becomes the *Plain* of Mars, by the swelling it produces in that part of the palm occupied by the Triangle [*vide* ¶¶ 380-1]; and as the

¶ 475. Construction of the mount and plain of ♂.

Plain of Mars is treated of in my remarks upon the triangle, but little notice need be taken of it here. The keynote of the whole question may be struck by bearing in mind that the *Mount* of Mars denotes *resistance*, whereas the *Plain* of Mars betrays action and aggression. This will be more fully demonstrated later on.

¶ 476.
Characteristics of the mount.

The main characteristics indicated by a development of the Mount of Mars are courage, calmness, *sang froid* in moments of emergency, resignation in misfortune, pride, resolution, resistance, and devotion, with a strong capacity to command.

¶ 477.
Its indications.

Well developed and not covered by lines or rays, this mount will counteract the evil influences of a short thumb by the calmness and resignation which it imparts to a character. Such a subject [especially if his thumb be large] possesses, to a marked extent, the capacity for keeping his temper. He will be magnanimous and generous to extravagance, loud of voice, and hot-blooded, his passions carrying him even to sensuality, unless counteracted by a strong phalanx of logic. His eloquence, if he possess that faculty, rare among subjects of this type, will be of the fascinating rather than of the emotional description. Spatulate fingers will give to this mount a love of show and self-glory.

¶ 478.
Marriage ♂ - ♀.

These subjects have always a great natural inclination to love, though they nearly always marry late in life, and marry women of the type of Venus [*vide* p. 224]. These two types seem to have a natural inclination for one another.[117]

¶ 479.
Aspect of the hand.

The hands to which these martial mounts belong are generally hard, the fingers large, especially at the

[117] "I know not how, but martial men are given to love. I think it is but as they are given to wine; for perils commonly ask to be paid in pleasures."—FRANCIS BACON'S "Essay on Love," 1612.

third phalanx, the will long, and the logic small, the hollow of the hand [Plain of Mars] rayed and lined.

An excessive development of this mount, [*i.e.*, a spreading of the mount into the palm, "the Plain of Mars,"] or a mass of lines upon the mount, will indicate brusquerie, fury, injustice of mind, insolence, violence, cruelty, blood-thirstiness, insult, and defiance of manner. Lines on the mount always denote hot temper. This excessive development generally betrays lasciviousness, and exaggeration in speech. ¶ 480. Excess and lines on the mount.

The Plain of Mars highly developed or covered with lines indicates a love of contest, struggle, and war, especially if the nails be short [*vide* ¶ 251] and a cross [*vide* ¶ 691] be found in the plain. This network of little lines in the Plain of Mars always indicates obstacles in the way of real good fortune. ¶ 481. Lines on the mount.

These hands of the excessive type have generally a feeble line of heart often joined to the line of head, the line of life red in colour, and the thumb short and clubbed. ¶ 482. Excessive type of ♂.

If the mount be completely absent, its absence denotes cowardliness and childishness. ¶ 483. Absence.

De Peruchio and Taisnier both assert that a line extending from the Mount of Mars to between the Mounts of Jupiter and Saturn, with little spots of the line of head, indicate deafness. I have never recognized the sign. ¶ 484. Tradition.

Combined with the Mount of Apollo, this mount becomes an indication of ardour and energy in art, force, perseverance, and truth in action. With the Mount of the Moon we get a love of navigation, or, if the rest of the hand is bad, [*vide* ¶ 408,] folly. Combined with the Mount of Venus we find a love of music and of dancing, sensuality, ardour, and jealousy in love. The combination of Mars and Mercury denotes movement and quickness of thought and speech, spontarvity, incredulity, and a love of ¶ 485. Combinations with other mounts. Apollo. Moon.

Venus.

Mercury.

CHEIROMANCY.

Saturn. argument, strife of words, and mockery. An equal development of the Mounts of Saturn and Mars gives cynicism, audacity of belief and opinion, and want of moral sense; we find, in fact, in this case, the energy of Mars rousing to action the usually latent evil qualities of Saturn.

MOUNT OF THE MOON.

§ 6. *The Mount of the Moon* [☽].

¶ 486.
Its attributes.

The attributes of this mount, when found predominant in a hand, are imagination, melancholy, chastity, poetry of soul, and a love of mystery, solitude, and silence, with a tendency to reverie and imagination. To it belongs also the domain of harmony in music, as opposed to the melody, which is the special attribute [as we shall see] of the Mount of Venus.

¶ 487.
Characteristics.

Such subjects are generally capricious and changeable, egoists, and inclined to be idle; their imagination often makes them hypochondriacal, and their abstraction often causes them to develop the faculty of presentiment, giving them intuition, prophetic instincts, and dreams. They are fond of voyages by reason of their restlessness, they are more mystic than religious, phlegmatic in habit, fantastic, and given to romance in matters of art and literature. They make generally the best rhymists, but they have no self-confidence, no perseverance, and no powers of expression in speech. They are much given to capricious marriages, which astonish their friends, from disparity of years, or something of the kind.

¶ 488.
Formation of the "lunar" hand.

These hands are generally swollen and soft, with short, smooth, and pointed fingers, and a short phalanx of logic. For the influence of the mount to be altogether good, it should be fuller at the base [near the wrist] than at the top [near the Mount of Mars] or in the centre. Excessive fulness in the exact centre

THE MOUNT OF THE MOON. 221

generally betrays some internal or intestinal weakness, whilst excessive fulness at the top indicates, as a rule, biliousness, goutiness, and a susceptibility to catarrh. Bad concomitant signs are a forking of the head line, [*vide* ¶ 261,] a low Mount of Mars, with the Mount of Apollo covered with a grille; then we find betrayed the vices of slander, debauchery, immodesty, insolence, and cowardice.

¶ 489. Hard hand.
The mount developed with a hard hand often betokens a dangerous activity and exercise of the imagination; with spatulate fingers this subject will be constantly forming projects and plans.

¶ 490. Effect of finger tips.
It may well be understood that a development of this mount emphasizes and harmonizes admirably with pointed fingers, but its development makes a square-fingered subject miserable by the constant turmoil and struggle between the realms of fact and fancy, *unless* there appear in the hand a good and well-traced line of Apollo, [*vide* p. 261,] which will give an artistic turn and instinct to the regularity of the square fingers. But if the fingers of the hand which bears this mount be very long, or *very* square, the inevitable result will be a perpetual discontent.

¶ 491. Suitable finge
A development of this hand should always [*vide* ¶¶ 131-2] be accompanied by short fingers, otherwise the detail indicated by the fingers will be constantly fretting the *laissez aller* instincts of the mount, or the morbid imagination of the mount will turn the detail of the fingers into a positive disease.

¶ 492. Excessive development
An excessive development of the Mount of the Moon will produce in a character unregulated caprice, wild imaginations, irritability, discontent, sadness, superstition, fanaticism, and error. Such subjects are intensely liable to suffer from headaches; and they take a morbid pleasure in painful thoughts and humiliating reflections.

¶ 493.
Long mount.

When the mount is not high, but very long, coming down to the base of the hand, and forming an angle with the wrist, it denotes a resigned and contemplative character, quite devoid of all strength, strength being shown by *thickness*, as opposed to weakness, which is indicated by *length* of the mount.

¶ 494.
Absence.

If the mount is absolutely absent, it betrays want of ideas and imagination, want of poetry of mind, and general drought of the intellect.

¶ 495.
Clairvoyance.

Highly developed with the "Croix Mystique," [*vide* p. 277] well traced in the hand, and pointed fingers, we find *invariably* a wonderful faculty of clairvoyance, which may be marvellously developed and cultivated.

¶ 496.
Idleness.

I have remarked above [¶ 487] that this mount indicates idleness; the idleness betrayed in a character by the development of this mount must not be confused with the idleness indicated by softness of the hands, [*vide* ¶ 203ᵃ,] the latter denoting idleness of the body, and slothfulness, as opposed to the idleness indicated by the former, which is that of the mind [reflection, etc.].

¶ 497.
Boundaries of the mount.

NOTE.—It sometimes occurs that there is a difficulty in determining the exact boundaries of the Mount of the Moon. It may generally be assumed that it joins the Mount of Mars at the extremity of the line of head, and is separated from the Triangle and the Plain of Mars by either the line of Saturn, or of Health, or by the Via Lasciva [which is rarely found in a hand, *vide* ¶ 640].

¶ 498.
Lines on the mount.

One line upon the mount betrays a vivid instinct, a curious vague presentiment of evils; many lines and rays on the mount denote visions, presentiments, prophetic dreams, and the like. Such subjects are much prone to folly and inconstancy. A single deep ray across the mount, with a small line crossing it, denotes gout or a gouty tendency.

A subject in whose hand is found a clear strong line from the Rascette to the middle of the mount [as at *b b*, in Plate X.] will be a complaining, fretful person.

¶ 499.
Line from the wrist.

A line extending in an arc from the Mount of Mercury to the Mount of the Moon, [as at *a a*, in Plate X.,] with more or less developed rays upon the mount, is an invariably sure sign of presentiments, prophetic instincts, and dreams.

¶ 500.
Connected with ☿ by curved line.

Horizontal lines traced upon the percussion at the side of the Mount of the Moon denote voyages. Such a travel line terminating with, or interrupted by, a star, indicates that the voyage will be a dangerous, if not a fatal one. If a travel line be so prolonged over the Mount of the Moon into the hand as to cut the line of head, making there a star, the subject will suddenly abandon his position and prospects in life, for the sake of a perilous voyage [*vide* ¶ 773].

¶ 501.
Horizontal lines. Voyages.

A star upon the mount connected by a small line with the line of life, is a prognostication of hysteria and madness [*c c*, in Plate X.] when it is accompanied by the other signs of dementia in a hand [*vide* ¶ 583].

¶ 502.
Star on the mount.

A straight line from the Mount of Mercury to that of the Moon betokens good fortune, arising from the imagination and guiding instinct developed in the mount.

¶ 503.
Connected with ☿ by straight line.

The mount much cross-barred indicates a condition of constant self-torment and worry, the cause of which will be shown by some strong development elsewhere in the hand, as, for instance, by a development of the line of heart, [*vide* ¶ 553,], which shows that the self-torment is from too much affection; or by a raying of the Mount of Jupiter, which shows ambition to be the disturbing element; or by a like condition of the Mount of Mercury, which indicates that the worries arise from business or commerce. This worrying tendency may, however, be

¶ 504.
Much cross-barred.

counteracted by very square fingers, or a long phalanx of logic; or it may be annulled by the resistance and resignation of a high Mount of Mars.

¶ 505.
Angle or crescent.

An angle on the mount [*d*, in Plate X.] indicates a great danger of drowning. A crescent in the same place is said to betoken the fatal influence of woman upon one's life. I have not come across these signs in practice.

¶ 506.
Combinations with other mounts.
Mercury.

If in a hand the Mounts of Moon and Mercury are equally developed, it is a sign of subtilty, changeability, and intuition in the deeper sciences, bringing, as their consequence, success and even celebrity. A

Venus.

like combination of the mount with that of Venus results in devotion of a romantic and fantastic kind, curiosity and *récherche* in affairs of the heart. In a bad hand such a combination will give caprice, eccentricity, and unnatural instincts in affairs of the heart. A

Saturn

combination with Saturn will give hypochondria and cowardice, egotism, slovenliness, and a tendency to indigestion. The constant attribute of the mount is imagination and fancy.

MOUNT OF VENUS.

§ 7. *The Mount of Venus* [♀].

¶ 507.
Its characteristics.

The main attributes of this mount, shown in a character by its prominence in the hand, are the possession of, and an admiration for, beauty, grace, melody in music, dancing, gallantry, tenderness, and benevolence, with a constant desire to please and to be appreciated. It is essentially the Mount of Melody, [*vide* ¶ 486,] and is, consequently, always to be found in the hands of those who are talented as singers. The attributes of this mount are the more feminine forms of beauty, as contrasted with the masculine forms of beauty, which are indicated by a prominence of the Mount of Jupiter.

¶ 508.
Its indications.

These subjects are great lovers of pleasure and society; they are fond of applause, but more from

THE MOUNT OF VENUS.

their love of giving pleasure to others than for its own sake. They hate any form of quarrel or strife, and are essentially gay, though they are less noisily gay, as a rule, than subjects of the type of Jupiter [*vide suprâ*]. Men of the type are often effeminate; all of them, however, have the talents of painting, poetry, and music, whether they have the perseverance to cultivate them or not.

A development of this mount will always mitigate and soften the harsh effects, or malignities, of any other mount.

¶ 509. Modifying effects of ♀.

The hands which usually accompany a development of this mount are fat and dimpled, the fingers smooth and rather short, the thumb also short. The *bad* influence of the type is betrayed by extreme softness, pointed fingers, the mount much cross-barred, the line of Mars indicated inside the line of life, and the Via Lasciva traced upon the palm.

¶ 510. The typical hand.

An excess of the mount will betray debauchery, effrontery, licence, inconstancy, vanity, flirtation, and levity.

¶ 511. Excess.

The absence of the mount betrays coldness, laziness, and dulness in matters of art. Without this mount developed to a certain extent, all the other passions become dry and selfish in their action.

¶ 512. Absence.

If the mount is completely devoid of lines, it indicates coldness, chastity, and, very often, a short life.

¶ 513. Very smooth.

A quantity of lines on the mount denotes always heat of passion and warmth of temperament. If there are but two or three strong lines traced upon the mount, they indicate ingratitude.

¶ 514. Lines on the mount.

A worn-out libertine has always this mount flat, but very much rayed, the Girdle of Venus [*vide* p. 268] being also traced in the hand, which indicates that the desire of the subject being beyond his powers, he constantly seeks for change and new excitement.

¶ 515. Debauchery.

¶ 516. *Connected with ☿.* A line extending from the mount to that of Mercury [*e e,* in Plate X.] is always a good sign, indicating good fortune and love resulting from one another.

¶ 517. *Line from wrist.* A line rising from the base of the hand into the mount is also a sign of good luck [*ff,* in Plate X.].

¶ 518. *Marriage lines.* Lines from the phalanx of logic to the line of life [*gg,* in Plate X.] are said by many authorities to indicate marriages; and if they are confused, they betray troubles and worries in love and marriage [*vide* ¶ 533].

¶ 519. *Islands.* Islands [*vide* ¶ 679] placed crosswise upon the mount [*j j,* in Plate X.] indicate advantageous opportunities of marriage which have been missed. These lost opportunities would have been all the more brilliant and desirable if the islands are connected with the Mount of Apollo [as at *k,* in Plate X.] by a line.[118]

¶ 520. *Other lines.* Three lines extending straight to the Mount of Jupiter denote liberality and happiness [*hh,* in Plate X.]. A deep line cutting into the triangle [*i i,* in Plate X.] betrays a tendency to asthma.

¶ 521. *The seven types.* NOTE.—It has been an almost invariable rule among cheirosophists to make these mounts the bases and distinguishing characteristics of seven clearly defined types, assigning to each a special physiognomy, phrenology, etc. I do not consider that this is expedient, for hands are already divided into seven far more practical and ordinary types cheirognomically [¶ 264], and in all my experience I have never found more than five or six subjects whose hands were dominated by one single pre-eminent mount [*vide* ¶ 398].

[118] Lunula in Monte Venereo adulterium monstrat; si montes Veneris atque Lunæ æque prominent, lunula in Monte Venereo obscenitatem prodit et crapulam; quin etiam si in tali manu Via Lasciva invenitur turpior atque apertior est lascivia.

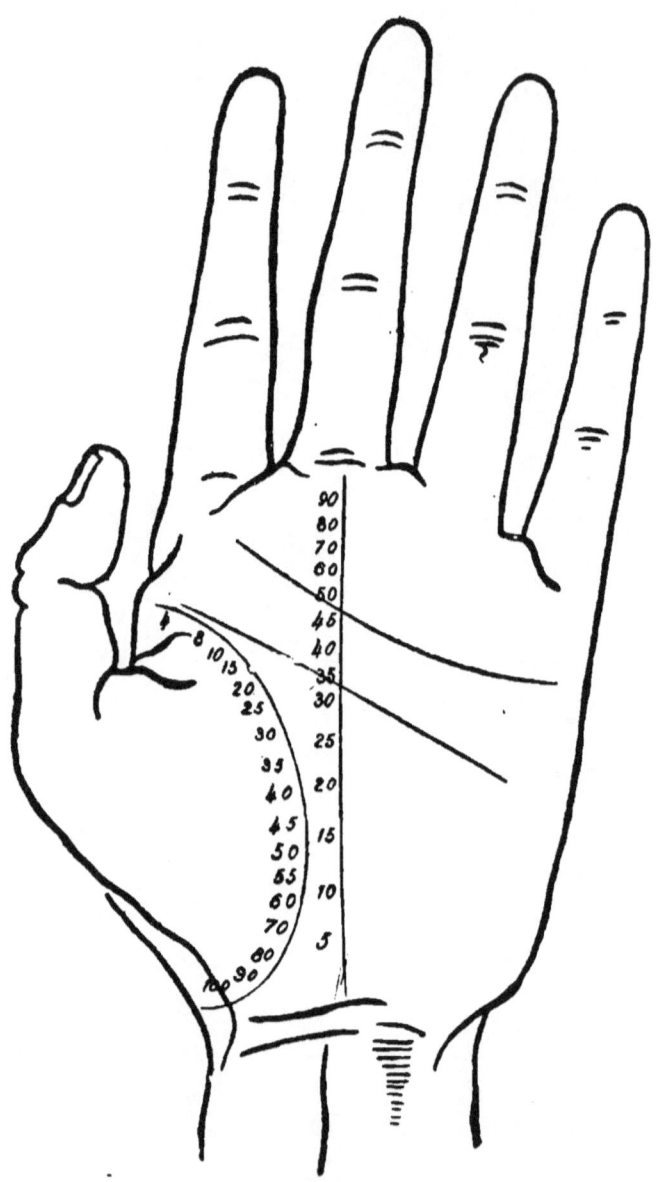

PLATE XI.—AGES UPON THE LINES OF LIFE AND FORTUNE.

SUB-SECTION IV.

THE LINES IN THE HAND.

THE LINES.

I SHALL consider and discuss each line in turn, according to its relative importance. The great difficulty about the consideration of the lines in the acquirement of the dogma of cheirosophy, is that the amount of details to be learnt by heart is apparently enormous. It is not, however, the case, as will be found when we reach the end of this sub-section, for, as a matter of fact, a complete knowledge of cheiromancy depends merely on a complete comprehension of the indications of the three principal lines—head, heart, and life. It is the aspect and *condition* of these lines, and the methods and causes of their disarrangements and subdivisions, which, properly observed, afford us all the information we can possibly require.

¶ 522.
Simplicity of the study.

§ 1. *The Line of Life.*

LINE OF LIFE.

This line should be long, completely encircling the ball of the thumb [Mount of Venus], strong, not too

¶ 523.
Proper conditions.

broad or too fine, without curvature, breakage, cross bars, or irregularities of any description. Thus marked in a hand, it denotes long life, good health, a good character and disposition.

¶ 524.
Evil aspects of the line.
Pale and broad, it indicates ill-health, bad instincts, and a feeble and envious character. Thick and red, it betrays violence and brutality of mind; chained, [*vide* fig. 6, Plate VIII.,] it indicates delicacy of constitution; thin and meagre in the centre, it indicates ill-health during a portion of the life; a spot terminating this thinness indicates sudden death. If it is of various thicknesses throughout its course it denotes a capricious and fickle temper.

¶ 525.
Age on the line.
Perhaps the most important consideration connected with this line is the determination of age. The line is divided up into periods of five and ten years, in the manner shown in Plate XI., and according as irregularities or breaks occur at any of these points, an illness or event whatsoever threatens the life at that age. [Thus, for instance, say a break occurs on a line of life at the point where you see the figure 40, you may predict an illness at that age or say the line ceases abruptly at the point 55, you may predict the death of the subject at that age.] It has often been objected to me that it is difficult to divide the line in a living hand from a diagram like Plate XI., owing to the difference in the size; but the difficulty ought not to exist, for the circumference of the Mount of Venus has only to be divided [mentally] into eighteen equal pa:ts, the points of division of which should be taken to represent the ages indicated on the diagram. A little experience will render this mental operation quite easy. The diagram given at p. 238 of Adrien Desbarrolles' smaller work on the science [*vide* Note [65], p. 67], has led many would-be cheirosophists wide of the mark, for the divisions are impracticable and incorrect, and his treatment of the other lines in the same way, by

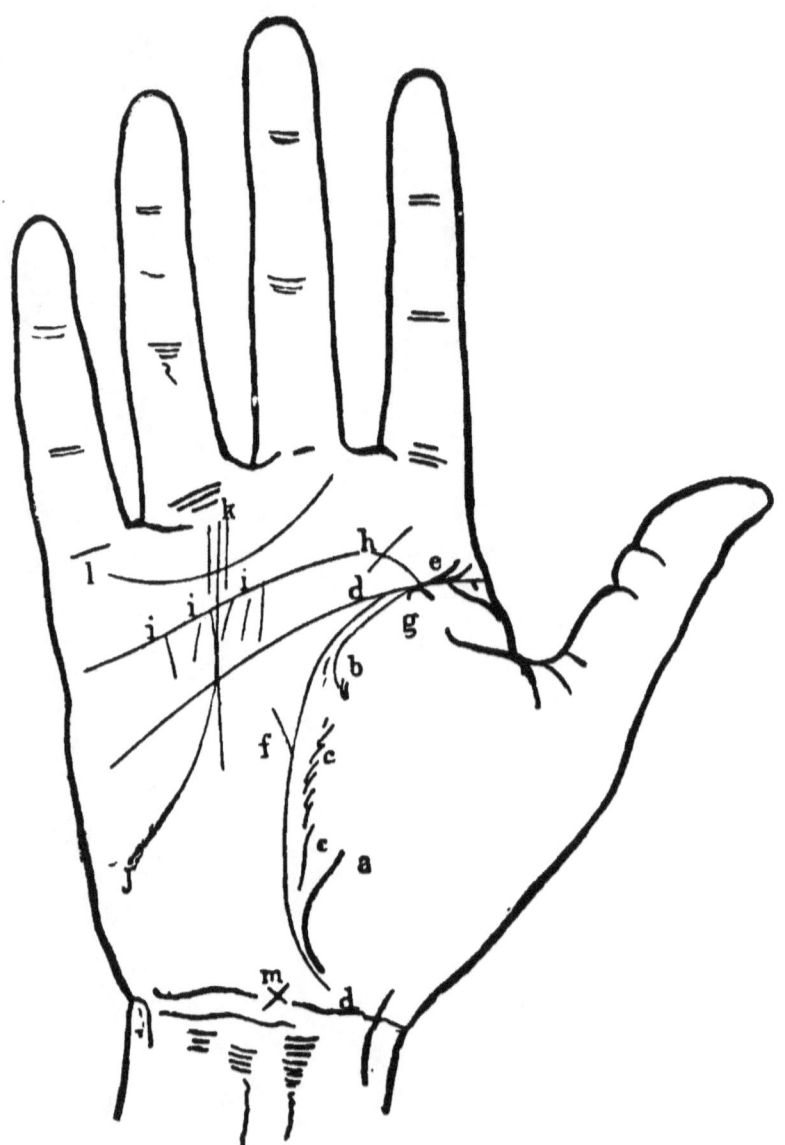

PLATE XII.—MODIFICATIONS OF THE PRINCIPAL LINES.

THE LINE OF LIFE. 233

rays from the line of life, is hopelessly and physically impossible. The method in which this division was obtained, and the astrological reasons and explanation thereof are given at length at p. 236 of the same work, but they are too lengthy and useless to transcribe here. He himself has recognized the fallacy of this method and subdivision, and hastens to correct them in the very first words of his later and larger work,[119] where he gives a diagram practically identical with Plate XI., as regards the divisions of the line.

¶ 526. Short line, short life.
The shorter the line the shorter the life, and from the point at which the line terminates in both hands may be predicted accurately the time at which death will supervene.

¶ 527. Breaks in the line.
A break in the line denotes always an illness. If the line is broken in both hands, there is great danger of death, *especially* if the lower branch of the break turn inwards towards the Mount of Venus [as at *a*, in Plate XII.], and the sign is repeated in *both* hands.

¶ 528. Necessity of corroborative signs in *both* hands.
And here I would digress to impress upon my readers a point of vital importance; that is, the absolute necessity to bear in mind that to be *certain* a sign *must* be repeated in *both* hands; and this applies particularly and especially to the indications of accident and disease upon the line of life. A break in one hand, and not in the other, betokens *only* a danger of illness; and in like manner, if in one hand the line stop short at 35, death cannot be predicted at that age, unless it also stop short at the same point in the other. For example, I saw a pair of hands some years since, in which the line ceased at 37 in one hand, and at 41 in the other. I told the subject that a fatal illness would attack him at 37, which would kill him at 41. He replied that he was then 39, and that a constitutional defect had asserted itself,

[119] "Mystères de la Main. Révélations complètes : Suite et fin." (Paris, 1874.)

in fact, at 37, and, as I had told him, he died two years afterwards. These things must be very carefully learnt before they are put into practice, for to make a deliberate statement like the above would be a brutal and dangerous thing to do, unless one spoke with *absolute* certainty.

¶ 529. Sudden death.
The line ceasing abruptly with a few little parallel lines as at *b*, in Plate XII., is an indication of sudden death. If the line is continually crossed by little cutting bars, it is an indication of continual, but not severe, illnesses.

¶ 530. Broken line.
Square.
If the line is broken up and laddered, as at *c c*, in Plate XII., it denotes a period of continued delicacy and ill-health. If it is broken inside a square, as at *a*, in Plate XIII., it indicates recovery from a serious illness; a square *always* denotes protection from some danger [*vide* ¶ 669]. A bar across the broken ends [as at *b*, Plate XIII.] also denotes a preservation from an illness.

¶ 531. Sister line.
Whatever may be the condition of the line, a sister line, as at *d' d*, Plate XII., will replace it and counteract the evil effects of the irregularities found on the main line, protect the subject against most of the dangers which assail him, and indicate a luxurious, comfortable existence. [Of the inner sister line, or line of Mars, we shall speak later on.]

¶ 532. Forked or tasselled.
The line should be free from forks and tassels throughout its course. Tasselled at its extremity, as at *c*, in Plate XIII., it indicates poverty and loss of money late in life, if not earlier. Forked at the commencement, as at *e*, Plate XII., it indicates vanity, indecision, and fantasy; but if the fork is very clear and simple, [not confused as in the figure,] it may in a good hand mean justice of soul and fidelity. In like manner, if instead of the tassel at *c*, Plate XIII., we find a plain fork, it points to overwork in old age resulting in poverty; it is, in fact, the first warning of the appearance of the tassel. A ray of the tassel going to the Mount of

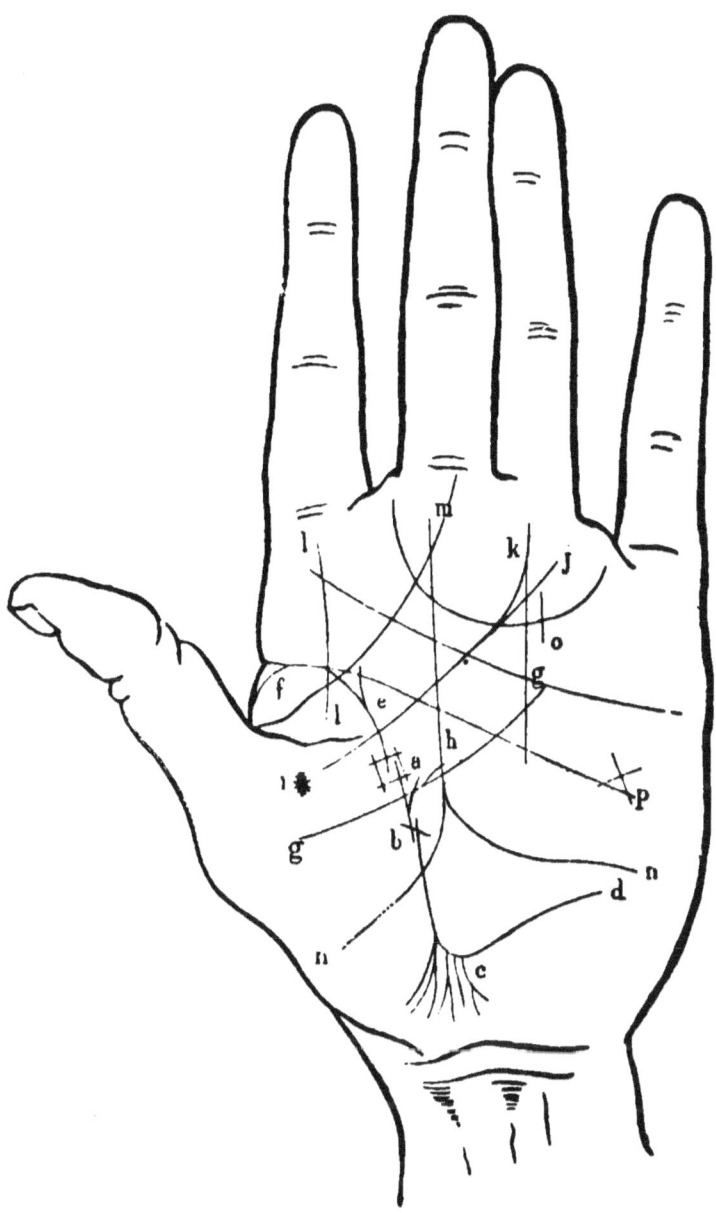

PLATE XIII.—MODIFICATIONS OF THE PRINCIPAL LINES.

THE LINE OF LIFE. 237

the Moon [as at *d*, Plate XIII.] shows a great danger of folly resulting from these troubles. A fork going to the line of head [as at *e*, Plate XIII.] equals faithfulness, but if it be at the side of the hand as at *f*, it is, on the contrary, a sign of inconstancy. A fork in the very centre of the line is a warning of diminished force, which *must* be attended to by a relaxation of the efforts, especially if the tassel appears at the base of the line, or the head is at all weak.

¶ 533.
Rays across the hand.
Worry lines.

Rays across the hand from the Mount of Venus [as in Plate XIV.] always denote worries and troubles. Across the line of fortune to a star in the triangle, they denote loss of money; continued to the line of head, as at *b*, a ray indicates a consequent loss of reason, or, at any rate, danger to the mental faculties. Cutting the line of Apollo, as at *c*, it betokens a worry or loss of money early in life, by reason of the ruin or misfortune of one's parents; if it starts from a star, as at *d*, it shows that the misfortune was caused by the death of a parent. The age at which these troubles occur is shown by the place at which the line of life is cut by the worry line. If the worry line terminates at a point or star upon the lines of head or heart [as from *f*, in Plate XIV.], or upon the Mount of Mars, it denotes that the worry has brought about an illness. If the line goes straight to the heart, as at *g g*, in Plate XIII., it indicates an unhappy love affair; if an island appear in the line, [*h*, Plate XIII.,] the consequences are likely to be, or have been, serious, if not shameful; a fork at the point where *g g* cuts the line of life, as in Plate XIII., indicates an unhappy marriage, or even a divorce. A worry line from a star in the mount [*i*, Plate XIII.] indicates quarrels with relations, ending in ruin if it goes up to the Mount of Apollo, as at *j*; but if it goes up and joins with the line of Apollo, as at *k*, it is a prognostic of good fortune arising therefrom. A line from the Mount of Venus, *just* cutting the line of life, as at *h*, in Plate XIV., indicates marriage at the age whereat the line is found.

¶ 534.
Rays cutting the line of life.
Rays across the hand just cutting the line, generally indicate an illness caused by the mount or line whence the ray takes its departure, at the age at which it occurs upon the line: thus, from the Line of Heart it means an illness caused by the heart; from the Line of Head an illness caused by the head or brain; from the Mount of Mars a danger brought about by passion, and so on.

¶ 535.
Ray up to ♃.
A ray ascending to the Mount of Jupiter, as at *l l*, in Plate XIII., betrays ambition, lofty aims, egoism, and success. These lines often appear in a hand quite suddenly.

¶ 536.
Spots on the line.
If a branch rise from a black spot on the line, it indicates that a disease has left a nervous complaint. Black spots always indicate diseases, and if they are very deep, they indicate sudden death. Among the older Cheiromants this was the indication of a murderer. These are more particularly treated of later on [*vide* ¶ 672].

¶ 537.
Ascending and descending branches.
Branches ascending from the line, as in Plate XV., denote ambition, and nearly always riches; if they ascend *through* the other lines, as at *a a a*, they indicate that the success is brought about by the personal merit of the subject. *Descending* branches, as at *b*, Plate XV., denote loss of health and wealth.

¶ 538.
Starting under ♃.
If instead of starting from the extreme outside of the hand, the line of life commences under the Mount of Jupiter, [say at *g*, Plate XII.,] it betrays great ambition, and is often a sign of great successes and honours.

¶ 539.
Joined to head and heart.
If the lines of life, head, and heart are *all* joined *together* at the commencement, it is a terrible sign of misfortune and violent death.

¶ 540.
Cross and branches.
A cross cut by branches of the line, as at *c*, Plate XV., betokens a mortal infirmity, with grave fear of death;

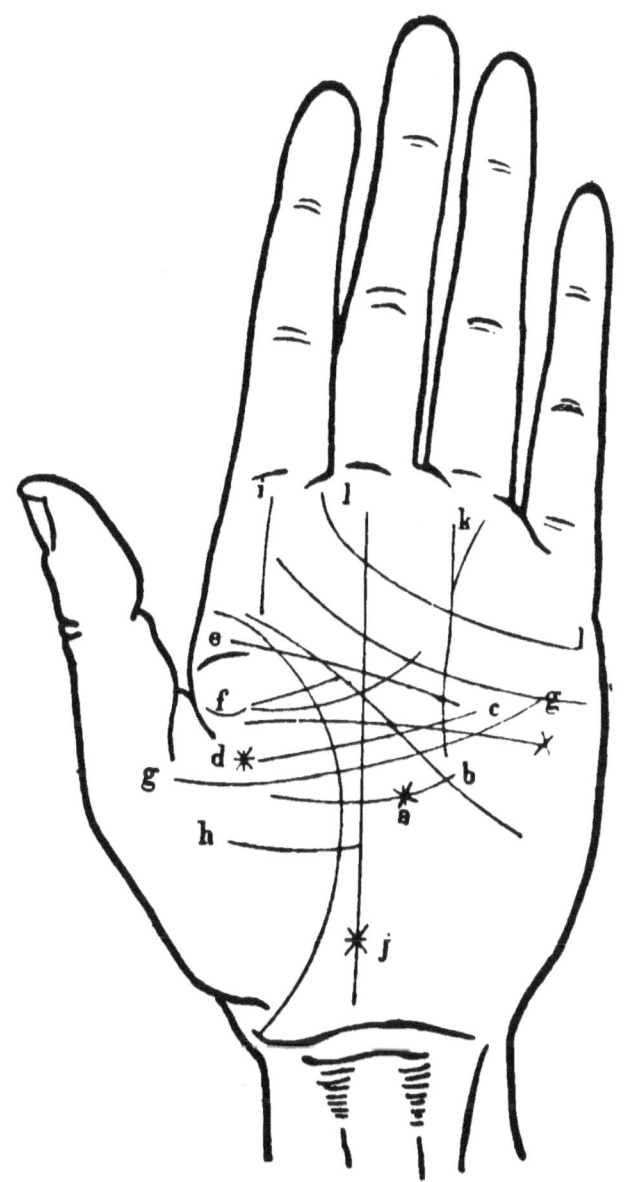

PLATE XIV.—MODIFICATIONS OF THE PRINCIPAL LINES.

a cross at the end of the line, as at *d*, denotes [if the line is otherwise clear] that the subject will suffer unmerited reverses in his old age. A cross at the commencement of the line indicates an accident in early life, especially if a point be also found on the line at the same place.

A line from the Mount of Mars cutting the line of life, as at *e e*, in Plate XV., indicates a wound.   ¶ 541. Line from ♂

A ray going direct from the line to the Mount of Apollo, denotes celebrity ; if it is indistinct, this celebrity is obstructed by some quality of the character, which must be sought for and guarded against.   ¶ 542. Ray to ☉.

Circles and spots upon the line were considered by the old cheiromants to indicate murder and blindness. I have seen the latter indication confirmed, but never the former.   ¶ 543. Circles and spots.

If the line, instead of being joined to the line of head, be separated, as at *f*, in Plate XV., it is a sign of folly and carelessness, of extreme self-reliance and foolhardiness in consequence, especially if the space be filled with a mesh of little lines, and the lines themselves be big and red.   ¶ 544. Separate from head.

If the line come out in a great circle into the palm of the hand, and reach, or end close to, the Mount of the Moon, it is a sign of long life. As I have said before, if a line have a break in it *and* a sister line, the latter mends it, as it were, and the only effect of the break is a delicacy during the period over which the break extends. If the broken end of the line join with the line of fortune, it is an indication that, at some time or other, the life has been in great danger, from which it has been protected by good luck.   ¶ 545. Curving out to mount of ☽.

Again, if the line appears to be short, an intense desire to live, supported by a strong phalanx of will and a good line of head, will often prolong it, the prolongation being marked on the hand by the appearance of sister lines or capillaries.   ¶ 546. Short line contracted.

¶ 547.
Close to thumb.

A line of life lying close to the thumb is a mark of sterility, especially if the lines of health and head are joined by a star [vide ¹²³, p. 267].

¶ 548.
Island

An island [vide ¶¶ 678-83] on the line denotes an illness during the period of its length, generally caused by some excess shown elsewhere on the hand. If the line of health is absent, the island denotes biliousness and indigestion; an island at the very commencement of the line betrays some mystery of birth, some fatality, or some hereditary disease.

LINE OF MARS.

§ § 1. *The Line of Mars.*

¶ 549.
Effects of the line.

In some hands we find inside the line of life, and running parallel and close to it, a second or sister line known in cheirosophy as the Line of Mars, or the Martial Line [vide Plate VII.]. Like all sister lines, it repairs and mitigates the effects of breaks in the main line; and it derives its name from the fact that it gives to soldiers great successes in arms, especially if it is clear and red in colour.

¶ 550.
Indications.

It gives, together with riches and prosperity, a great heat and violence to the passions, which with this line, if uncontrolled, are apt to become brutish. Its influence lasts throughout the period during which it follows the line of life; and De Peruchio says that there is always a love affair at the age at which it begins.

LINE OF HEART.

§ 2. *The Line of Heart.*

¶ 551.
Proper aspects.

This line should be neat, well coloured, and extending from the Mount of Jupiter to the outside of the hand under the Mount of Mercury, not broad and pale, or thick and red, but well traced, and of a good normal colour; such a condition of the line indicates a good heart, an affectionate disposition, with an equable temper and good health.

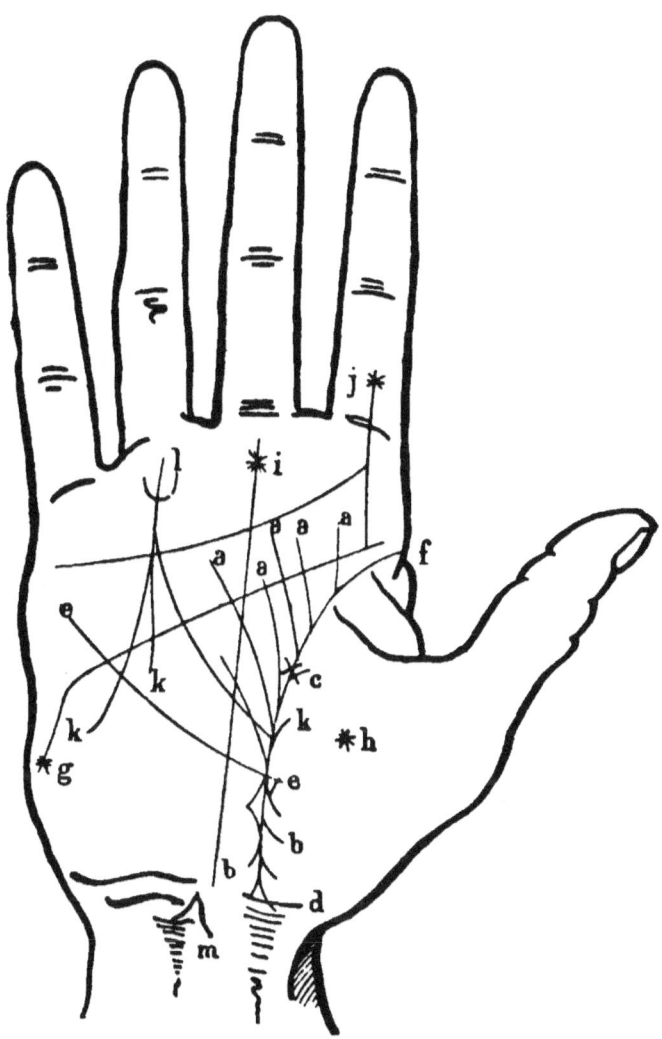

PLATE XV.—MODIFICATIONS OF THE PRINCIPAL LINES.

The strength of the affection is in proportion to the length of the line; if the line, instead of beginning at the Mount of Jupiter, begins upon the Mount of Saturn, the subject will be more sensual than Platonic in his affections.

¶ 552. Length of the line.

Traced right across the hand, [from side to side,] it indicates an excess of affection which produces jealousy and suffering in consequence thereof, especially if the Mount of the Moon is high.

¶ 553. Excess.

If it is chained in its formation, the subject is an inveterate flirt, and, unless the rest of the hand be very strong, will be much subject to palpitations of the heart.

¶ 554. Chained.

Bright red in colour, the line denotes violence in affairs of the heart, and, on the other hand, a pale line, broad and chained, betrays a cold-blooded *roué*, if not a condition of heart utterly *blasé*. A livid or yellow colour betrays subjection to liver complaints.

¶ 555. Colour.

The line should be close underneath, well up to the bases of the mounts; a line which lies close to that of the head throughout its length, betrays evil instincts, avarice, envy, hypocrisy, and duplicity.

¶ 556. Position.

A line of heart which begins quite suddenly without branches or rays beneath the Mount of Saturn, foreshadows a short life and a sudden death. If the line is very thin and runs right across the hand, it indicates cruelty even to murderous instincts.

¶ 557. Commencing under ♄.

If the line, instead of terminating on the Mounts of Jupiter or Saturn, seems to disappear between the first and second fingers, it betokens a long life of unremitting labour.

¶ 558. Between first and second fingers.

If to a large line of heart a subject add the Girdle of Venus [*vide* § 7] and a high Mount of the Moon, he will be a victim to the most unreasoning jealousy.

¶ 559. Girdle of ♀ and ☽.

If in a hand there be found *no* line of heart, it is an unfailing sign of treachery, hypocrisy, and the worst

¶ 560. Absence of the line.

instincts, and, unless the line of health be very good, the subject will be liable to heart disease, and runs a grave danger of a sudden early death.

¶ 560°.
Breaks in the line.
A line which is much broken up denotes inconstancy, and often these subjects are woman-haters. A single break shows a feebleness of the heart, and the cause of that feebleness may always be found in some excess or evil development of a mount— fatality shown by a development of the Mount of Saturn; foolishness shown by an equal development of the Mounts of Saturn and Apollo; pride shown by the Mount of Apollo; folly or avarice shown by the Mount of Mercury.

¶ 561.
Many li.tle lines.
A quantity of little lines cutting across the line diagonally indicate many misfortunes of the heart, arising originally from weakness of the heart or liver.

¶ 562.
Forked.
The line dividing at the end and going in three branches to the Mount of Jupiter, is a most fortunate sign, indicating riches and good luck. Any forking of the line which sends a branch on to the Mount of Jupiter is good; even if the branch goes to between the fingers of Jupiter and Saturn, this betokens still good fortune, but of a more quiet and undisturbing description. But a forking which sends one ray upon the Mount of Jupiter and the other upon the Mount of Saturn, betrays errors and failures in the search after success, and fanaticism in religion.

¶ 563.
Bare line.
If the line is quite bare under the finger of Jupiter at its commencement, there is great danger of poverty; a similar bareness at the percussion indicates sterility; the fork under Jupiter gives also to the subject energy and enthusiasm in love. A line quite bare of branches throughout its length indicates dryness of heart and want of affection.

¶ 564.
Touching ♃.
If the line touch the base of the finger of Jupiter, the subject will be unsuccessful in all his undertakings, unless the line of fortune be exceptionally good.

A mark like a deep scar across the line betrays a tendency to apoplexy; red spots or holes in the line denote wounds either physical or moral. White marks on the line denote conquests in love; a point on the line means grief of the heart, and, according to its position, you can tell by whom it was caused, thus:—Under the Mount of Apollo the cause was an artist, or a celebrity—*i.e.*, the grief is connected with art or ambition; under the Mount of Mercury the grief is caused by a man of science, a lawyer, or a doctor.

¶ 565. Marks on the line

If the line curl round the first finger, it is a sign of a marvellous faculty for occultism and the possession of high occult powers.

¶ 566. Curled round first finger.

Joined to the line of head under the Mounts of Jupiter or Saturn, is a sign of a great danger threatening the life, and of sudden and violent death, if the sign is repeated in both hands. If the line turn down on to the line of head, with a ray across it, as at *h*, in Plate XII., it is a sign of a miserable marriage, or deep griefs of the heart.

¶ 567. Joined to head.

A ray from the line of life to the Mount of Saturn, reaching to the base of the finger, [as at *m*, in Plate XIII.,] is a very bad sign in a woman's hand, immeasurably and even fatally increasing the dangers of maternity.

¶ 568. Ray to ♄.

Lines from the quadrangle [*vide* p. 288] to the line of heart, as at *i i i*, in Plate XII., denote aptitude for science, curiosity, research, and versatility, which often culminates in uselessness.

¶ 569. Lines from the quadrangle.

A curved line from the line of heart to the Mount of the Moon [stopping *abruptly* at the line of heart (*vide* ¶ 500)], as at *j*, in Plate XII., denotes murderous tendencies and instincts.

¶ 570. Curved line to ☽.

§ 3. *The Line of Head.*

LINE OF HEAD.

This line should be joined to the line of life at its immediate commencement, and leaving it directly should trace a strong ray across the hand to the top of the

¶ 571. Proper aspects.

Mount of the Moon, clear and well coloured, without ramifications or forking, uninterrupted and regular; such a formation indicates good sense, clear judgment, cleverness, and strength of will.

¶ 572.
Evil conditions and aspects.
Pale and broad, it indicates feebleness or want of intellect. Short—*i.e.*, reaching only to the Plain of Mars,—it betrays weak ideas and weak will. [Stopping under the Mount of Saturn, it foreshadows an early sudden death.] Chained, it betrays a want of fixity of ideas and vacillation of mind. Long and very thin, it denotes treachery and infidelity. Of unequal thickness, twisted, and badly coloured, it betrays a feeble liver and want of spirit; such subjects are always avaricious.

¶ 573.
Length of the line.
A long line of head gives domination to a character —*i.e.*, domination of self as opposed to the domination of others, indicated by a large thumb. A long line of head in a many-rayed and lined hand gives great self-control and coolness in danger and difficulties, and the strength of the head [shown by the long line] causes the subject to reason out and utilize the intuitive powers and instinctive promptings indicated by the multiplicity of rays and lines in the hand.

¶ 574.
Excessive length.
Very long and straight,—*i.e.*, cutting the entire hand in a straight line from the line of life to the percussion,—it indicates excess of reasoning habits, over-calculation, and over-economy, denoting avarice and meanness.

¶ 575.
Modifications.
The excessive economy [avarice] of this long line may be greatly modified by a softness of the hand or a high development of the Mounts of Jupiter or of Apollo.

¶ 576.
Starting under ♄.
If instead of joining the line of life at its commencement it only leaves it under the Mount of Saturn, it is a sure indication that the education has been acquired and the brain developed late in

THE LINE OF HEAD.

life ; or, if the line of life is short, and the line of head also, it foreshadows a grave danger of sudden death. A like commencement, the line reaching across to the Mount of Mars, the line of heart being thin and small, indicates struggles and misfortunes arising from infirmities of temper or errors of calculation, unless the line of fortune is exceptionally good. Such a subject will often appear benevolent, but his benevolence will generally be found to be only of a nature which gives pleasure to himself, and is usually more theoretical than practical.

The line must lie at a good regular distance from that of the heart; lying close up to it throughout its length, it betrays weakness and palpitations of the organ. *¶ 577. Position.*

Remember that an extremely good line of head may so influence the whole hand as to dominate other evil signs which may there be found, especially if the Mount of Mars be also high; such a combination gives to a subject energy, circumspection, constancy, coolness, and a power of resistance which goes a long way towards combating any evil or weak tendencies which may be found in his hand. *¶ 578. Influence of good line.*

If the line stops abruptly under the Mount of Saturn it forewarns of a cessation of the intelligence, or [with other signs] death in early youth; stopping similarly under the finger of Apollo, it betrays inconstancy in the ideas and a want of order in the mind. *¶ 579. Stopped under ♄ or ☉.*

If, though visible, it appears joined to the line of life for some way before leaving it to go across the hand, it indicates timidity and want of confidence, which give dulness and apathy to the life, and which are with difficulty overcome. When this sign appears in an otherwise clever hand, the most strenuous efforts should be made to counteract this want of self-reliance, which is so serious an obstacle to *¶ 580. Joined to life at commencement.*

success. Joined to the line of life in a really *strong* and clever hand, the indication will be of caution and circumspection.

¶ 581.
Thin at centre.

Thin in the centre for a short space, the line indicates a nervous illness, neuralgia, or some kindred disease.

¶ 582.
Separate from life.

Separated from the line of life at its commencement and going well across the hand, it indicates intelligence, self-reliance, and spontaneity, [*vide* ¶ 544,] and, with a long thumb, ambition. Separate from the line of life, and short or weak, it betrays carelessness, fantasy, jealousy, and deceit; often these subjects have bad sight. Separated thus, but connected by branches or ramifications, it indicates evil temper and capriciousness; connected by a cross, it betrays domestic troubles and discomforts. Even in a good hand there is danger in this sign of brusquerie, and a too great promptitude of decision which often leads to error. With the Mounts of Saturn or Mars prominently developed, it is a sign of great audacity or imprudence, but it is a useful prognostic [within limits] for public characters or actors, giving them enthusiasm and boldness of manner in public, and the gift of eloquence by reason of their self-confidence. A *long* line thus separated will give want of tact and discrimination, and an impulsive manner of speech, which is often inconvenient, and sometimes wounds.[120]

¶ 583.
Declining to the mount of ☽.

If the line, instead of going straight across the hand to the base of the Mount of Mars or to the top of the Mount of the Moon, trace an oblique course to a termination *on* the Mount of the Moon, it is

[120] Linea capitalis ita dispincta si curta, et ad Lunam decidens in manu quâ linea mensalis catenæ similis et truncatus, Cinctus Venereus, quâ in Monte Venereo cancelli apparent et absente linea. Appollinaris pollex curtus est, et crassæ tertiæ digitorum vertebræ: impudicum hominem haud dubio demonstrat. Si omnia hæc signa in manu videntur, impudicitiam certe atque aperte declarant.

THE LINE OF HEAD.

a sign of idealism, imagination, and want of instinct of real life. If it comes very low upon the mount it leads to mysticism and folly, even culminating in madness if the line of health is cut by it in both hands. In an otherwise fairly strong hand this declension upon the Mount of the Moon gives poetry and a love of the mystic or occult sciences, superstition, and an inclination to spiritualism. Such a formation, if the Mount of the Moon is rayed, generally gives a talent for literature. The line of head coming low upon the Mount of the Moon to a star, as at *g*, in Plate XV., with stars on the Mounts of Venus and Saturn, as at *h* and *i*, and a weak line of heart, are terribly certain signs of hereditary madness. This extreme obliquity of the line always indicates a *danger* of madness, and these concomitant signs [*vide* ¶ 419] prove it to be hereditary, and probably unavoidable.

Again, if instead of going across the hand it turns up towards one of the mounts, it will show that the thoughts are entirely taken up by the qualities belonging to the respective mounts; thus turning up to the Mount of Mercury commerce will be the prevailing instinct, and will bring good fortune; or, turning towards the Mount of Apollo, a desire for reputation will be the continual thought. If it points between the fingers of Apollo and Mercury, the signification is of success in art brought by scientific treatment. If the line go right up on to the mount it will denote a folly of the quality—thus, for instance, ending on Mercury it will denote occultism and deceit; on Apollo, the mania of art; and on Saturn, the mania of religion.

¶ 584. Turning up to a mount.

Any turning up of the line of head towards that of the heart denotes a weak mind, which lets his heart and his passions domineer over his reason; if it *touch* the line of heart it is a prognostic of early death. If it cut through the line of heart and end

¶ 585. Turning up to, or cutting, heart

upon the Mount of Saturn, it foreshadows death from a wound to the head. I have seen this sign verified in two terrible instances. If it turn up to the line of heart and confound itself with it obliquely, it foreshadows a fatal affection, which runs a great risk of terminating in madness.

¶ 586.
Turning back.

Turning back towards the thumb, the line of head denotes intense egotism and misfortune in consequence thereof.

¶ 587.
Breaks in the line.

A break in the line of head nearly always indicates an injury to the head. Broken under the finger of Saturn, and the broken ends overlapping, as at *a*, in Plate XVI., the prognostic is especially certain, but in a bad hand it is said to be a sign of the scaffold, or, at any rate, of the loss of a member, *even* if the sign appear in one hand only. Much broken up it is a sign of headaches and general weakness of the head, resulting in loss of memory and want of continuity in the ideas. Such a breaking up will rob a long phalanx of will of much of its power, and long fingers of much of their spirit of minutiæ. If with this shattered line of the head we find in the Plain of Mars a cross, the rays terminating in points or spots and short nails, it is a grave warning of a tendency to epilepsy.

Cross in ♂.

¶ 588.
Split and sister line.

Split throughout its length is a strengthening sign if other indications of madness appear in the hand, but if the line is distinctly *double* [*i.e.*, if it is accompanied by a sister line] it is a sure sign of good fortune and inheritances.

¶ 589.
Forked at the end.

If the line is forked at the end, with one of the "prongs" descending upon the Mount of the Moon, [as at *b*, in Plate XVI.,] we have a certain indication of lying, hypocrisy, and deceit. Such a man, even with a good hand, will be a clever sophist, never off his guard, ready at all times with [if necessary] an ingenious rearrangement of facts to suit the needs of

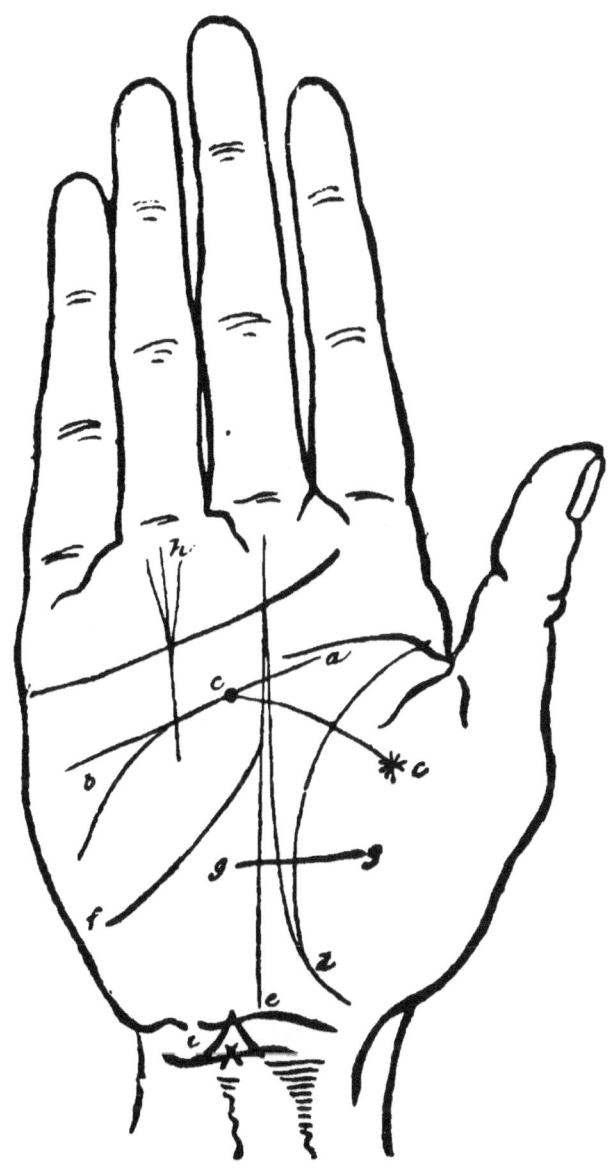

PLATE XVI.—MODIFICATIONS OF THE PRINCIPAL LINES.

the immediate present. This forking has somewhat the effect of short nails, giving to a subject a love of controversy and argument. If the rays or "prongs" of the fork are so long that one extends right across the hand, and the other comes well down to the rascette, it has the dual effect of a long line of head, and of a line of head which descends far upon the Mount of the Moon, giving at once poetry and realism—*i.e.*, a capability of making a practical use of poetic inspirations. A good line of Apollo gives *great* talent to a forked line of head, from its power of seeing all round a subject, and of considering it from all points. If one ray of the fork go up to touch the line of heart, and the other descends upon the Mount of the Moon, it betrays the sacrifice of all things to an affection, and if with this sign the line of Saturn or fortune stops short at the line of heart, it denotes that this infatuation has brought ruin with it. The two signs are nearly always concomitant.

¶ 590. Cut by lines.

Cut by a multitude of little lines, the line of head indicates a short life, with many illnesses and headaches. If the little cross lines are confined to the middle of the line of head it is a sign of dishonesty.

¶ 591. Cross.

A cross in the middle of the line is a foreshadowing of near approaching death, or of a mortal wound if the line is also broken at this point.

¶ 592. Points on the line.

Red points indicate wounds; white ones indicate discoveries in science or inventions. Black points, ailments according to the mount most developed in the hand. Thus with the Mount of Saturn, toothaches; with the Mount of Venus, deafness; with the Mount of Apollo, diseases of the eyes [especially if a star appear at the junction of the finger of Apollo and the palm]. These points are often connected with similar spots on the line of life by rays or lines, which enable us to pronounce with certainty the ages at which the subject has suffered from these maladies.

¶ 593.
Knotted.

A knotting up of the line betrays an impulse to murder, which, if the knot is pale, is past, but which, if the knot is deep red, is to come.

¶ 594.
Capillaries.

Capillary lines [*vide* fig. 9, Plate VIII.] on the line of head are a sign of a well-ordered mind and a good disposition.

¶ 595.
Island.

An island in the line of head is an indication of acutely sensitive nerves.

¶ 596.
Star.

A star upon the line is generally a sign of a very bad wound, bringing danger of folly with it.[131]

¶ 597.
Star on ♀.

If a line be found connecting a star on the Mount of Venus with a spot on the line of head, [as at *c c*, in Plate XVI.,] it indicates a deeply-rooted and ever-remembered disappointment in love.

¶ 598.
Line to ♃.

If a line extend from the line of head to the root of the finger of Jupiter, [as at *i*, in Plate XIV.,] it indicates intense pride and vanity which is easily wounded; if it ends at a star upon the finger, [as at *j*, in Plate XV.,] it is a sign of extreme good luck; but if it ends at the same place by a cross, the luck will be, on the contrary, extremely bad. This little line, joined by the line of Saturn or fortune, indicates vanity, reaching even to folly.

LINE OF FORTUNE.

§ 4. *The Line of Saturn, or Fortune.*

¶ 599.
Its points of departure and their indications.

The line of Saturn, or fortune, has three principal points of departure for its base: it may start from the line of life, as at *d*, in Plate XVI.; from the *rascette*, as at *e;* or from the Mount of the Moon, as at *f*. Starting from the line of life, the line of fortune indicates that the luck in life is the result of one's own personal merit. If it starts from the wrist, or rascette, the fortune will be very good, especially if it trace a fine strong furrow

---

[131] Stella in Lineâ Capitali, ubi cum Hepaticâ jungitur, prætendit feminis periculosam luçinam: si præclare stella signatur, sterilitatem indicat.

on the Mount of Saturn; in the same direction, but commencing higher up from a point in the Plain of Mars, we get an indication of a painful, troubled life, especially if the line penetrate [as it often does] into the finger. If the line start from the Mount of the Moon, it shows [if it goes straight to the Mount of Saturn] that the fortune is, to a great extent, derived from the caprice of the opposite sex. If from the Mount of the Moon the line goes to that of the heart, and, confounding itself therein, goes on up to the Mount of Jupiter, it is an infallible sign of a rich and fortunate marriage. You must guard against confounding a chance line from the Mount of the Moon to the line of Saturn with the line of Saturn starting from that Mount. If [besides the line of Saturn, as at *e*, in Plate XVI.] we have another line starting as at *f*, in Plate XVI., and *cutting* instead of joining the line of Saturn, it betrays the fatal effects of imagination, culminating possibly in weakness, or evil to the mental capacity. Starting from the very base of the Mount of the Moon, and ending on the Mount of Saturn, is an indication of prediction and clairvoyance.

¶ 600. Termination of the line.

Instead of going to the Mount of Saturn, the line may go up to some other mount, in which cases it will have special significations; thus, going to the Mount of Mercury, we get fortune in commerce, eloquence, and science; going to the Mount of Apollo, we get fortune from art or wealth; going to the Mount of Jupiter, we find satisfied pride, and the attainment of the objects of our ambition.

¶ 601. Length of the line.

If the line, instead of stopping on the mount, goes right up to the second joint of the finger, we have the indication of very great fortune, which will be either very good or very bad, according to the concomitant signs. Thus, with a good hand, this is a first-rate sign; but with a deep red line on the mount, and a star on the first phalanx of the finger, we have the indication of the worst possible fortune, ending in a violent death, probably on the scaffold. The line should just extend from the top of the rascette to the centre of the Mount of Saturn ; reaching to the jointure of the finger and palm, or penetrating into the rascette is a bad sign, being a sure indication of misery.

Stopped at lines of head or heart. Starting from the rascette, and stopped at the line of heart, indicates a misfortune arising from a disappointment in love; or, in a weak hand, heart-disease. Similarly stopped at the line of head, the misfortune will arise from an error of calculation, or from an illness of the head.

¶ 602.
Starting short.

If it only *start* from the line of head it denotes labour, pain, and ill-health, unless the line of head is very good, when it will be an indication of fortune acquired late in life by the intelligence of the subject. Shorter still,—*i.e.*, from the quadrangle to the Mount of Saturn,—the indications are still more unfortunate, being of great sorrows, and even of imprisonment. The evil prognostications of a line which goes into the third phalanx of the finger of Saturn may be averted by the presence of a square [*vide* ¶ 669] on the mount.

¶ 603.
Broken in quadrangle.

If the line is stopped in the quadrangle, and then starts again at the line of heart, ending its course upon the mount, it denotes that though the luck will be obstructed and retarded, it will not be permanently spoilt, and the position in life will not be lost ; and this is especially certain if a good line of Apollo be found in the hand.

¶ 604.
Age on the line of fate.

And this brings us to the indications of age on the line of Saturn. The line starts from its base, and on it [as in Plate XI.] one can tell by its breaks, and so on, approximately the ages at which events have occurred in a life : it must, however, be premised that these indications are not anything like as sure as those of the line of life. From the base of the line to the line

THE LINE OF SATURN, OR FORTUNE.

of head we have thirty years, from the line of head to that of the heart we find the events of the life between thirty and forty-five years, and thence to the top of the line takes us to the end of the life. Thus, for instance, if you see a gap, or break, in the line from the line of head to just below the line of heart, you can predict misfortunes between the ages of thirty and forty; and a connecting line will generally indicate the nature and cause of the ill-luck. Also it will often be found that in the right hand a misfortune will be marked on the line of Saturn, the exact *date* of which will be marked by a point on the line in the left hand.

The indications found upon the line of Saturn often explain and elucidate indications only dimly or vaguely traced upon the line of life, or in the rest of the hand. ¶ 605. Explantions by ♃.

A perfectly straight line, with branches going upwards from its two sides, indicates a gradual progress from poverty to riches. Twisted at the base, and straight at the top, indicates early misfortunes, followed by good luck. Straightness, and good colour, from the line of heart upwards, always betokens good fortune in old age, with invention in science, and a talent for such pursuits as horticulture, agriculture, construction, and architecture. Split and twisted, the line of Saturn indicates ill-health from an abuse of pleasure. A twisted condition of the line always denotes quarrels, and a very good and well-traced line of Saturn will annul the evil indications of a badly-formed line of life. ¶ 606. Conditions of the line. Twisted

Split.

A broken-up and ragged condition of the line betrays an inconstancy and changeability of fortune Breaks in the line in the Plain of Mars, denote physical and moral struggles. Even, however, if it is broken up, it may be replaced by a very good development of the Mount of Saturn, or a favourable ¶ 607. Breaks in the line.

aspect of the Mount of Mars; and to the worst luck a high Mount of the Moon will give a calm and resignation which rob it of much of its evil indication. A strong, irregular line of fortune, in a much-rayed and lined hand, betrays a constant irritability, and a super-sensitive condition of mind. A well-traced line of Saturn always gives a long life; broken up at the base is an indication of misery in early life, up to the age [vide ¶ 604] at which the breaking up ceases. If it ends in a star on the mount, it foreshadows great misfortune, following great good luck; in a good hand this sign generally means that the misfortune is caused by the fault of others, *generally* of one's relations. For the line of Saturn to be lucky, there must be explanatory points in the hand for the luck to come from, and to find these is one of the most important tasks of the cheirosophist.

¶ 608. Cut by lines. Cut by a multitude of little lines on the mount, we can safely foretell misfortunes late in life, after a long period of good luck. Cut by a line parting from the Mount of Venus, it denotes conjugal misery, or misfortune caused by a woman [*g g*, in Plate XVI.].

¶ 609. Absence. If the line is simply absent from a hand, it denotes an insignificant life, which takes things as they come, meeting with neither particularly good nor particularly bad fortune.

¶ 610. Forked. Forked, with one ray going to the Mount of Venus and the other to the Mount of the Moon, [as at *n n*, in Plate XIII.,] we find a strife for success, directed by the wildest imagination, and spurred on by love. If the line go well up, as in Plate XIII., the ambition will be successful, after much struggle; but if the main line is broken or malformed, the necessary intrigues and caprices caused by the formation of the line will result in inevitable misfortune.

¶ 611. Crosses. Any cross upon the line indicates a change of position or of prospects in life at the age indicated by the position of the cross upon the line [as in Plate XI.]. In the very centre of the line it is always a misfortune, and the cause of it may nearly always be found upon the lines of head or life, showing the misfortune to arise from error or miscalculation, or from illnesses or the loss of friends.

¶ 612. Stars.

A star at the base of the line [as at *j*, in Plate XIV.] denotes a loss of fortune, brought by the parents of the subject in early youth; if there be also a star on the Mount of Venus, [as at *h*, in Plate XV.,] the immediate cause is the early death of a parent.

¶ 613. Island.

An island on the line betrays, *almost invariably*, a conjugal infidelity; a star accompanying the island betokens a great misfortune arising therefrom. At the very base of a line, an island indicates a mystery connected with the birth of the subject, and with this sign, an extreme malformation of the line will betray illegitimacy. In a really good hand, an island on the line of Saturn indicates a hopeless, untold passion; with a star and a cross on the Mount of Jupiter, the island will show that the passion has been for a celebrated or exalted person.[122]

§ 5. *The Line of Apollo, or Brilliancy.*

LINE OF APOLLO.

¶ 614. Position in the hand.

The line of brilliancy may start either from the Line of Life, the Plain of Mars, or the Mount of the Moon, as at *k k k*, in Plate XV. Whenever it is present, it denotes glory, celebrity, art, wealth, merit, or success; its best aspect is when it is neat and straight, making a clear cut upon the Mount of Apollo, signifying celebrity in art, and consequent riches, with a capacity for enjoying and making the best of them. Clearly marked, the line also denotes that the subject is under

---

[122] Insula in Lineâ Saturniâ adulterium monstrat: et longa insula hominem nonnullos annos pro insulæ longitudine in adulterio vixisse declarat [*vide* ¶ 789, et Pl. XX.]. Cum longâ pollice et bonâ capitale lineâ etiam hoc signum faustum est.

the favour or influence of the great; it gives him, also, the calmness of natural talent, and the contentment of self-approbation.

¶ 615. Necessity in lucky hand. It is necessary that this line exist in a really lucky hand to make its good fortune absolute; a good Line of Saturn will be seriously compromised by the absence of this line.

¶ 616. With ♃ and ☿. With the Mounts of Jupiter and Mercury developed, this line is a certain indication of wealth, and such a subject will become celebrated by his fortune, dignity, and merit, no less than by his talents and scientific capacities.

¶ 617. Twisted fingers. Twisted fingers, or a hollow palm, are very bad signs with this line; for they show that the influences of the line are guided in an evil direction, and that the talents betokened by it are used for the attainment of bad ends.

¶ 618. With long head and ☉. With a long line of head, and a long finger of Apollo, the tendencies of the line will be material, the ambition and talents being turned towards the attainment of riches.

¶ 619. Proper aspects of the line. The line, to have all its highest artistic significations, should be well coloured; pale, it denotes that the subject is not actively artistic, but has merely the instincts of art, loving things that are brilliant and beautiful. In these respects the indications are the same as those of a high Mount of Apollo *without* the line; such a formation also gives a love of the beautiful *without* production, the mount giving the instincts, and the line giving the talents, of art.

¶ 620. Absence. Absence of the line from a hand indicates want of success in projects and undertakings which would [if successful] lead to glory and success.

Broken up. Much broken up it indicates a Jack-of-all-trades and an eccentricity in art which renders it of little avail to the owner.

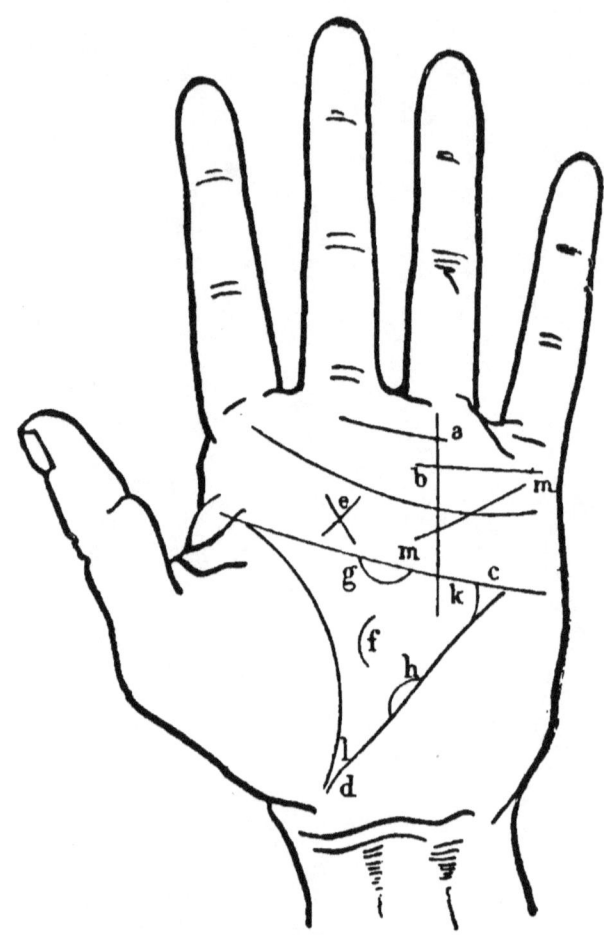

PLATE XVII.—MODIFICATIONS OF THE PRINCIPAL LINES, THE QUADRANGLE AND THE TRIANGLE.

Many little lines upon the mount point generally to an excess of artistic instinct, which generally falls by its own weight, and comes to nothing; it is much better to have only one line on the mount, unless all are equally clear and well traced. With two or three lines, a subject will often follow two or three different branches of art, without succeeding particularly in any one.

¶ 621. Lined mount of ☉.

If the line is confused and split up in the quadrangle, but clear above, we find misfortunes, having, however, good terminations. Any sign upon the line of Apollo in the quadrangle must be carefully observed, for they always denote worries, and are generally connected by a worry-line [*vide* ¶ 533] with the line of life and Mount of Venus, showing the times at which they occurred.

¶ 622. Signs in the quadrangle.

If the line is equally divided on the mount, as at *k*, in Plate XIV., we find an equal balancing of two instincts, which ends in a nullity in the matter of art. Divided into a curved trident, as at *l*, in Plate XV., it is a sure indication of vast unrealized desires of wealth; if, however, the line divides into a pointed trident from the line of heart, as at *h*, in Plate XVI., we can safely announce future glory, riches, and celebrity arising from personal merit; and if, instead of being joined at the heart, the three lines rise parallel and identical, as at *k*, in Plate XII., tracing three fine troughs on the mount, we have these same indications intensified and made yet more certain.

¶ 623. Divided on the mount.

Cross-lines on the mount are obstacles which stand in the way of artistic success, very often arising from the envy and malignity of others.

¶ 624. Cross lines.

Cut by a line coming from the Mount of Saturn, as at *a*, in Plate XVII., poverty will stand in the way of complete success. Similarly cut by a line coming from the Mount of Mercury, as at *b*, in Plate XVII., the success

¶ 625. Line from ♄ and ☿.

and good fortune will be marred and prevented by inconstancy and changeableness of spirit.

¶ 626. Star. A star on the mount is a good sign, indicating success and good luck, arising from the favour of others and the help of friends.

¶ 627. Cross. A cross upon the mount *close to* the line, or even touching it, denotes instinct of religion and piety.

¶ 628. Spot. A black spot at the junction of the lines of heart and of Apollo betrays a great danger, if not an imminent peril of blindness.

LINE OF HEALTH.

§ 6. *The Line of Liver, or Health.*

¶ 629. Position in the hand. The position which the liver line [line of health, or *Lignea hepatica*] occupies in the hand may be seen by looking at the Map of the Hand, Plate VII., but it will strike the cheirosophist, after very little experience, that this line fully developed in a hand is comparatively scarce, and the reason of this I take to be the comparatively unhealthy lives which the majority of peolpe live nowadays. I have seen this line in a fresh young hand, beautifully traced and as clear as any of the others, and watching the hand have seen the line break up and practically disappear in the course of a few years.

¶ 630. Proper aspects. Long, clearly traced, and well coloured and proportioned, the line denotes good health, gaiety, a clear conscience, and success in life. If it is lengthened up to the upper part of the palm it is a sign that the good health of the subject will last well into old age. A thoroughly good line of health will counteract the evil effects of a poor line of life, being an indication of good digestion, which will always prove a powerful agent in prolonging life.

¶ 631. Base of the line. The line should not be joined to that of the life at its base, but just separated, as at *d*, in Plate XVII.—this will indicate long life; joined at this point with the line of life, it is a sure indication of weakness of the heart.

This line completely absent from a hand will render the subject vivacious in conversation, agile and quick in manner.

¶ 632.
Absent.

If the line is thick and blunt, it is a sign of sickness in old age; if it is very straight and thin, it is a sign of rigidity of spirit and manner. Red at the upper end, it betrays a tendency to headaches; thin and red in the centre, it is a sign of fever; red at the lower end is a sign of a weak heart; thus it will be seen that any unevenness of colour in this line is bad. Very red throughout its length indicates brutality and pride. A twisted and wavy liver line is a sign of biliousness, and very often of dishonesty, of which it is, at any rate, a strong confirmatory indication. Much broken or cut into, the line will betray a weak digestion.

¶ 633.
Evil conditions of the line.
Colour.

Forked at the top so as to make a triangle with the line of head, [as at *c*, in Plate XVII.,] it gives a great love of honours and power combined, *always* with a marvellous aptitude and capacity for occult sciences.[123]

¶ 634.
Forked at top.

A *coming* sickness marks itself on this line by a little deep cross-line; a past sickness marks only the life or head lines, leaving merely a gap in the line of health.

¶ 635.
Sicknesses.

The line of health making a good clear triangle with the lines of head and of fortune, we find a subject very clever at natural magic, electro-biology, and the like, a great student of nature and of natural phenomena, with a high faculty of tuition, sometimes accompanied by second sight.

¶ 636.
Clear triangle

The line traced across upon the Mount of the Moon is a sure sign of caprice and of change in the course of the life of the subject.

¶ 637.
Going to ☽.

[123] Stella apud juncturam Capitalis Lineæ cum Hepaticâ sterilem feminam demonstrat, præsertim si Linea Vitalis prope pollicem currit.

¶ 638.
Island.
A long island at the base of the line denotes a somnambulist.

¶ 639.
Sister line.
A sister line to this liver line indicates strong and unprincipled avarice.

CEPHALIC.
§§ 1. *The Cephalic Line, or Via Lasciva.*

¶ 640.
Position.
This line is rare; it is often confounded with the line of health, and is still more often regarded as a sister line to the liver line, but it is quite a separate line from itself, appearing only *conjointly with* the line of head, though it diverges considerably from it in the manner shown in the Map of the Hand.

¶ 641.
Indications of the line.
It generally betokens cunning, and often faithlessness, especially if twisted, though these indications are considerably modified the more distinct it be from the line of health.

¶ 642.
Length.
It gives ardour and fervour to the passions, and reaching up to the Mount of Mercury, it indicates constant good luck arising from eloquence and pure talent.

¶ 643.
Stars.

Joined to ☉.
Stars on the line generally betoken riches, but often they betray serious troubles and struggles in front of, and accompanying them. Joined by a ray to the Line of Apollo, the line is a sure indication of wealth.

¶ 644.
Joined to ♀.
It used to be customary to look upon this line [as its ancient name denotes] as a sign of lasciviousness, but this indication only belongs to it if it runs across into the Mount of Venus.

GIRDLE OF VENUS.
§ 7. *The Girdle of Venus.*

¶ 645.
General character.
This line, fortunately not universal, may be taken, as a whole, to be a *bad* sign in any hand, indicating a tendency to debauchery, which it is extremely difficult to conquer.

¶ 646.
In a good hand.
To a good hand, however, this line will expend itself by giving energy and ardour in every under-

taking entered into by the subject, and this favourable influence of the line is the more certain if it is clear, neat, and going off upon the Mount of Mercury, as at *ll*, in Plate XIV. To a good hand this will give merely love of pleasure and energy therein.

¶ 647. Effect of the line.

It generally makes a subject hysterical and nervous, with a great tendency towards spiritualism and sorcery, accompanied by a more or less chronic state of melancholy and depression. There is also very frequently a talent for and a love of literature, and lyric poetry.[124]

¶ 648. Cutting fate or ☉.

If the lines of fortune or of Apollo are cut by the Girdle of Venus, so as apparently to shatter them in two at this point on the mount, it is a sign of obstacles to the success and misfortunes, probably the result of excessive passion, or ardour in the pursuit of pleasure.

¶ 649. On to ☿.

Coming up on to the Mount of Mercury, as at *l*, in Plate XII., the subject will add to all the other evil indications of the line the vices of lying and theft.

¶ 650. Cut upon ☉.

Cut upon the Mount of Apollo by a short deep line, [as at *o*, in Plate XIII.,] it is a sign of loss of fortune, caused by dissipation and debauchery.

¶ 651. Hysteria.

Crossed by a quantity of little lines, it is a sure sign of a hysterical nature, especially if the Mount of Venus, or of the Moon, or both, are highly developed.

¶ 652. Retrospect.

We have now considered the principal lines, and discussed them with considerable minuteness; a careful retrospect will show the reader that [as I said in ¶ 370] the indications of the lines are easily found

---

[124] Cinctus Venereus truncatus et lacer, cinædum indicat et ad pravissimam et libidinosissimam lasciviam proclivem. Quin etiam certior indicatio, si Cinctus Venereus duplex vel triplex est, et quo planior eo deterior. Homines, in quorum manibus hæc signa videntur, se st—t, quo flagitio mentes perdunt vel inutiles faciunt. Linea secans Cingulum Venereum in Monte Saturnio mortem per cædem apud meretrices portendit. Stella in Cinctu venereum morbum semper indicat.

by examining their condition with reference to the mounts and the other lines of the palm, each mount or line having its peculiar significations and effects, and bringing them to bear upon the *other* mounts and lines and the qualities indicated by them, by juxta-position or connection with them by means of lines, rays, or signs.

¶ 653.
Chance lines.
Often, however, we find lines in a hand which cannot be accounted for by any of the foregoing rules, and these [which are called "chance lines"] are made the special subject of a future chapter. The signs found in the palm, though they have frequently been adverted to in the previous sub-section, will be our next consideration, with reference to their special and individual significations.

[*Vide* SS. VIII.]

SUB-SECTION V.

(Plate IX.)

THE SIGNS IN THE PALM.

SIGNS.

GREAT attention must be paid to the signs which are found very frequently upon, or close to, the mounts and lines of the hands, for they very greatly modify and alter the recognized significations of the mounts or lines, and generally carry with them an indication entirely their own.

¶ 654.
Their importance.

§ 1. *The Star.*

STARS.

A star, [fig. 10, Plate IX.] wherever it appears, is generally the indication of some event we cannot possibly control; it is generally a danger, and always something unavoidable. Whether, however, it is good or bad, depends of course upon the aspect of the lines, particularly of the line of fortune. This, however, is fixed—that a star, wherever it is found, always means *something*, and what that something is, be it the task of the cheirosophist to discover.

¶ 655.
Indication.

On the Mount of Jupiter it signifies gratified ambition, good luck, honour, love, and success. With a cross on this mount it indicates a happy marriage with some one of brilliant antecedents or high position.

¶ 656.
On ♃.

¶ 657.
On ♄.

On the Mount of Saturn it indicates a great fatality, generally a very bad one, indicating, with corroborative signs, probable murder, and in a criminal or otherwise very bad hand a probability of death upon the scaffold.

¶ 658.
On ☉.

On the Mount of Apollo, with no line of Apollo [*vide* p. 261] in the hand, it betokens wealth without happiness, and celebrity after a hazardous struggle for it. *With* the line of brilliancy it denotes excessive celebrity, as the combined result of labour and talent; with several lines also on the mount it is a sure indication of wealth.

¶ 659.
On ☿ or ♂.

On the Mount of Mercury it betrays dishonesty and theft. On the Mount of Mars violence leading to homicide.

¶ 660.
On ☽.

On the Mount of the Moon it indicates hypocrisy and dissimulation, with misfortune resulting from excess of the imagination. The old cheiromants looked upon this as a warning of death by drowning, and stated that combined with a high mount invaded by the line of head, it indicated suicide by drowning.

¶ 661.
On base of ♀.

On the *base* of the Mount of Venus it indicates a misfortune brought about by the influence of women.

¶ 662.
On the phalanges of the fingers.
On ♄.

On the first [or outer] phalanx of any finger [but especially of that of Saturn] a star indicates either strange good luck or else folly. On the third [or lowest] phalanx of the finger of Saturn, a star warns the subject of a danger of assassination, and if at this point it is joined by the line of Saturn, a disgraceful death is almost inevitable, resulting, as a rule, from the vices shown elsewhere in the hand.

¶ 663.
On the thumb.

On the base of the phalanx of logic of the thumb,—in fact, on the junction of the phalanx of logic and the Mount of Venus,—it points to a misfortune connected with a woman, probably indicating an unhappy marriage, which will be the curse of the subject's whole existence, unless the Mount of Jupiter be developed, in which case there is a probability that the subject will get over it.

A star on a voyage line [*vide* ¶¶ 773 and 501] indicates with certainty death by drowning. ¶ 664. On a voyage line.

If a star be found in the centre of the quadrangle, the subject, though true and honest as the day, will be the absolute plaything of woman, a trait which will result in a misfortune, from which, however, he will recover in time. ¶ 665. In the quadrangle.

Thus it will be seen that a star is almost the most important sign to seek for in a hand. ¶ 666. Its importance

§ 2. *The Square.* SQUARE.

The appearance of a square [fig. 11, Plate IX.] on the hand always denotes power or energy of the qualities indicated by the mount or line on which it is found. It is a sign of good sense, and of cold, unimpassioned justice. ¶ 667. Effect.

It may either appear as a neat quadrangular figure, traced as if with a punch, or it may be formed of the [apparently] accidental crossing of principal and chance lines. It will often appear enclosing a bad sign, from the effects of which it entirely protects the subject. ¶ 668. Appearance and position.

Wherever it is found it always denotes protection; thus round a break in the line of life [*vide* ¶¶ 530 and 602] it betokens recovery from that illness; or on the line of Saturn, it will protect the subject from the evil effects of a badly-formed line, or of bad signs found thereon. ¶ 669. Protection.

A star on the Mount of Saturn surrounded by a square denotes an escape from assassination; a square with red points at the corners denotes a preservation from fire. ¶ 670. With star on ♄.

The square has one evil signification—that is, when it is on the Mount of Venus, close to the line of life; ¶ 671. On ♀.

18

under these circumstances it is a warning of imprisonment of some sort or another.

§ 3. *The Spot.*

Spot

¶ 672.
Its indication.

A spot, [figs. 1 and 12,] wherever found and of whatever colour, always denotes a malady; placed upon a line, it is nearly always the mark of a wound; on the line of head it denotes a blow to the head, and consequent folly.

¶ 673.
Colour.

A white spot on the line of heart denotes a conquest in love; a white spot on the line of head points to a scientific discovery. A red spot is the sign of a wound; a black or blue spot is the sign of a disease, generally of a *nervous* character. The white spot is the only comparatively harmless one.

§ 4. *The Circle.*

Circle.

¶ 674.
On ☉.

The circle [fig. 13, Plate IX.] is a comparatively rare sign, which has only one good signification—that is, when it appears on the Mount of Apollo, where it indicates glory and success.

¶ 675.
On ☽.

On the Mount of the Moon it denotes danger of death by drowning; on any other mount it gives a dangerous brilliancy.

¶ 676.
On the lines.

On any line it is bad, denoting always an injury to the organ or quality represented. Thus, on the line of heart, it betrays weakness of the heart, and on the line of head it forewarns a subject of blindness.

§ 5. *The Island.*

Island.

¶ 677.
Its distinctness.

The island [fig. 14] should perhaps more properly have been noticed in treating of the lines generally; but it is a sign so distinct from any ordinary formation of the line, that I have thought it best to consider it in this place as a sign proper.

An island means always one of two things; either it is the mark of something disgraceful, or else it betrays an hereditary evil. It is the more often an hereditary malady of the line, as, for instance, on the line of head it will show an hereditary weakness of the head, or on the line of heart it betrays an hereditary heart disease, and so on.

¶ 678. Its indications

As for the disgraceful indications of the island, it should be taken to mean more properly that the *chance, i.e.,* the temptation, will occur; but a long line of head and a strong phalanx of will on the thumb will always annul the most evilly-disposed island.

¶ 679. Evil indications.

On the line of heart it means in a good hand heart disease, or, in a bad one, adultery.

¶ 680. On line of heart.

On the line of head, if it occur on the Plain of Mars, it shows a murderous tendency; if *beyond* the Plain of Mars, it betrays evil thoughts. On a good hand it will merely indicate hereditary head weakness.

¶ 681. On line of head.

On the line of liver or health it betrays a tendency to theft or dishonesty; in a good hand a weak digestion, or an intestinal complaint.

¶ 682. On liver line.

On the line of life an island indicates some mystery connected with the birth.

¶ 683. On life.

§ 6. *The Triangle.*

TRIANGLE.

The triangle [fig. 15] always denotes aptitude for science, and may be formed either neatly and by itself, or by the [apparently] chance coincidence of three lines.

¶ 684. Its indication.

On the Mount of Jupiter it indicates diplomatic ability. On the Mount of Saturn it betrays aptitude for occult sciences and necromancy, a sign which becomes very sinister and evil if there be also a star on the third phalanx of this finger. On the Mount of Apollo a triangle indicates science in art; on the Mount of Mercury, talent in politics; on the Mount

¶ 685. On the mounts

of Mars, science in war; on the Mount of the Moon, wisdom in mysticism; and on the Mount of Venus, calculation and interest in love.

§ 7. *The Cross.*

CROSS.

¶ 686.
Its effect.

The cross [fig. 16] is seldom a favourable sign, unless it is *very* clearly and well marked, when by accentuating the qualities of the mount or line, it may have a good signification. It nearly *always* indicates a change of position.

¶ 687.
On ♃.

Its one undoubtedly *good* signification is when it appears on the Mount of Jupiter, when it denotes a happy marriage, especially if the lines of Saturn or of Apollo start from the Mount of the Moon. [*Vide* ¶ 435.]

¶ 688.
On ♄.

On the Mount of Saturn it denotes error and fanaticism in religion or occult science, leading to the more evil forms of mysticism.

¶ 689.
On ☉.

On the Mount of Apollo it betrays errors of judgment in art, unless there be also a fine line of Apollo, which will give to the cross the significations of wealth.

¶ 690.
On ☿.

On the Mount of Mercury it indicates dishonesty, and even theft.

¶ 691.
On ♂.

On the Mount of Mars it denotes danger arising from quarrelsomeness and obstinacy.

¶ 692.
On ☽.

A cross on the Mount of the Moon will indicate a liar, and a man who deceives even himself if it is large; but if it is small, it will merely indicate reverie and mysticism.

¶ 693.
On ♀.

On the Mount of Venus it denotes a single and a fatal love, unless another cross appear on the Mount of Jupiter [*vide* ¶ 435] to render the union happy.

¶ 694.
At the base of the hand.

At the bottom of the hand, near the line of life,— *i.e.*, in the lower angle of the triangle,—a cross denotes a struggle, ending in a change of position in life, which is the more radical according as the cross is more or less clearly marked at this point.

§ § 1. *The " Croix Mystique."*

CROIX MYSTIQUE.

This is a sign so entirely by itself that I devote a separate discussion to it. It is found traced with more or less distinctness in the quadrangle beneath the finger of Saturn.

¶ 695. Its position.

It always gives to a subject mysticism, superstition, and occultism, or, with a very good hand, religion. If it is very large it betrays exaggerated superstition, bigotry, and hallucination.

¶ 696. Its indications

If it is clearly traced in both hands, it betrays folly arising from the excessive influence of the principal mount; thus, with Jupiter developed, over-ambition; with Saturn, misanthropy; with Apollo, extreme vanity or miserliness; and with Venus, erotomania.

¶ 697. In both hands.

If the "Croix Mystique" is joined to the line of Saturn, it foretells good fortune arising from religion.

¶ 698. Joined to ♄

If it is displaced, so as to lie, as it were, between the Mounts of Mars and of the Moon, [as at *p*, in Plate XIII.,] it indicates a changeability of disposition, which will lead to good fortune.

¶ 699. Displacement

§ 8. *The Grille.*

GRILLE.

The grille [fig. 17] is generally the indication of obstacles, and of the faults of a mount whereon it is found. But if there be no mount particularly elevated in the hand, it will so emphasize a mount, if it is found upon one, as to make *it* the principal mount and keynote of the interpretation of the hand.

¶ 700. Its indications.

On the Mount of Jupiter it indicates superstition, egoism, pride, and the spirit of domination.

¶ 701. On the mounts.
♃.

On the Mount of Saturn it foretells misfortune and want of luck.

♄.

CHEIROMANCY.

☉.
On the Mount of Apollo it betrays folly and vanity, and a great desire of glory, joined to impotence and error.

☿.
On the Mount of Mercury it tells of a serious tendency towards theft, cunning, and dishonesty.

♂.
On the Mount of Mars it forewarns a violent death, or, at any rate, some great danger thereof.

☽.
A grille on the Mount of the Moon indicates sadness, restlessness, discontent, and a morbid imagination.

¶ 702.
On ☽ with a lined hand.
If on a hand which is much covered with lines [*vide* ¶ 421] it shows a constant movement and state of excitement. If there be a star on the Mount of Saturn this sign tells of the wildest exaltation, nervous spasms, and continual anxieties and disquietude. With a well-traced line of Apollo and a grille on the Mount of the Moon we find poetry, and great talent for lyrics and literature.

¶ 703.
On ♀.
The grille on the Mount of Venus is often a bad sign, denoting lasciviousness and morbid curiosity, especially with the Girdle of Venus traced in the hand. With a strong phalanx of will and a long line of head and the line of Apollo, or brilliancy, this sign merely results in a nervous excitement, which is in no way pernicious or evil in its effects, giving a refinement and daintiness to the passions.

¶ 704.
Modifying signs.
A strong phalanx of will, with a good line of head and of Apollo, will *always* greatly modify the sinister effects of the grille, *excepting* when it is found on the Mounts of Jupiter or Saturn, when it is practically irremediable.

PLANETARY SYMBOLS.

§ 9. *The Signs of the Planets.*

¶ 705.
Effects.
Besides the above comparatively ordinary signs we find in some instances [though such instances are *excessively* rare] the actual sign of a planet actually traced on a mount. As a rule, when this occurs the

rest of the hand is perfectly plain, the whole force of the character being concentrated in the quality indicated by the "precipitation" of the planetary sign. As these are so intensely rare, I will give three examples only which I have had the fortune actually to see myself.

The sign of Mercury [☿] traced upon the Mount of Jupiter gives great administrative talent and noble eloquence. The sign of the Moon [☽] on the Mount of Jupiter leads to intense mysticism and error. The sign of Mercury on the Mount of Apollo gives great celebrity and eloquence in science.

¶ 706. Combinations

A mount sometimes also, instead of being high or rayed, has its own sign traced upon it; thus ♃ on Jupiter, ♄ on Saturn, ☉ on Apollo, ☿ on Mercury, ♂ on Mars, ☽ on the Moon, and ♀ on Venus. These signs, of course, intensify the qualities of the mounts to an extremely marked and extraordinary extent.

¶ 707. On its own mount.

SUB-SECTION VI.

THE SIGNS UPON THE FINGERS.

SIGNS UPON THE FINGERS.

IN the preceding sub-section we have dealt only with the signs found upon the palm of the hand. We have also to consider the lines and signs which find themselves traced upon the fingers, which signs have also their special significations.

¶ 708.
Lines on the first phalanx.
Lines on the first phalanx of a finger always denote a weakness or failing of the quality of the finger. If the lines are twisted and confused they foreshadow danger to the subject from the qualities of the finger. A single deep ray on the first phalanx of a finger indicates an idealism or folly connected with the quality.

¶ 709.
Lines connecting the phalanges.
Lines from the first into the second phalanges unite, as it were, the worlds of idealism and reason, [*vide* ¶ 146,] causing the subject to mix a certain amount of reason with all the promptings of his

imagination. In the same way lines connecting the second and third phalanges unite reason and matter, and the subject will always set about his worldly affairs in a reasonable and sensible manner.

One short line, sharply traced on each phalanx of each finger, is a prognostic of sudden death. ¶ 710. One line on each phalanx.

Lines running the entire length of the fingers give energy and ardour to the qualities of the finger; cross lines, however, are obstacles in the way of the proper development of the characteristics of the finger. ¶ 711. Lines all along the fingers.

§ 1. *Signs on the First Finger, or Index.* First Finger.

A line extending from the mount, through the third phalanx into the second, gives a character in which reason and thought is mingled with audacity. ¶ 712. Line to second phalanx.

Cross lines on the third phalanx indicate inheritances, according to the older cheirosophists; on the second phalanx they denote envy and falsehood. Lines across the tips of the fingers denote general debility, and if they extend all the way from one side of the nail round the ball of the finger to the other side, they foreshadow wounds to the head. ¶ 713. Cross lines.

A pair of crosses on the second phalanx are a sign of the friendship of great men. ¶ 714. Crosses.

A star on the first phalanx indicates great good fortune; a star on the second phalanx indicates mischief and boldness, unless it is connected with the first phalanx by a line, in which case it becomes a sign of modesty. A star on the third phalanx is a sign of inchastity. ¶ 715. Stars.

A crescent upon the first phalanx is a sure sign of imprudence, which may bring about very grave results. ¶ 716. Crescent.

§ 2. Signs on the Second or Middle Finger.

Second Finger.

¶ 717. *Into third phalanx.* — A line from the Mount of Saturn across the third phalanx of the finger indicates prosperity in arms; if it is oblique it foretells death in battle.

¶ 718. *Lines.* — Many lines just penetrating into the mount denote cruelty; if they go the whole length of the finger they indicate melancholy; or, if they are very parallel and equal, they denote success in mining operations. If the lines are confined to the first phalanx they denote avarice. Twisted lines on the third phalanx denote ill luck.

¶ 719. *Triangle.* — A triangle on the third phalanx indicates mischief and ill luck.

¶ 720. *Cross.* — A cross in the same place indicates sterility in a female hand.

¶ 721. *Star.* — A star on the first phalanx indicates great misfortune, and if it is on the *side* of the finger it betrays a probability of death, which will, however, be in a just cause.

§ 3. Signs on the Third, or Ring Finger.

Third Finger.

¶ 722. *Lines.* — A single line running the entire length of the finger is a sure indication of great renown. Many lines are a sign of losses, probably occasioned by women.

¶ 723. *Lines on the third phalanx.* — Straight lines on the third phalanx indicate prudence and happiness. Turning to one side of the finger they indicate great success, but not success accompanied by wealth. If the lines on the third phalanx penetrate on to the mount they indicate good fortune, accompanied by loquacity and often by arrogance.

¶ 724. *From third into second.* — A line extending from the third phalanx into the second is a sign of goodness and cleverness, accompanied by good fortune. Cross lines placed upon this phalanx indicate difficulties in the way which will have to be surmounted.

A crescent on the third phalanx signifies unhappiness, and a cross at the same place signifies extravagance. ¶ 725. Crescent.

§ 4. *Signs on the Fourth or Little Finger.*

FOURTH FINGER.

A line throughout the length of this finger is a signification of success in science and uprightness of mind; three lines similarly running right down the finger are a sign of research in chimærical and impossible sciences. ¶ 726. Lines.

Deep lines on the first phalanx denote weakness of constitution; a cross on the same place is significant of poverty and *consequent* celibacy. ¶ 727. On the first phalanx.

Lines on the second phalanx are an indication of research in occult sciences. If they are confused and coarse they betray inchastity. ¶ 728. On the second phalanx.

A line from the third into the second phalanx indicates eloquence and consequent success. If the line is twisted it gives great sharpness and cunning in defence of self. If this line start from the mount it is a still surer sign of prosperity and success. ¶ 729. From third to second phalanx.

One thick line, a scar, or a cross on the third phalanx, betrays a tendency to theft. A star on the same phalanx denotes eloquence. ¶ 730. Star, cross, or star.

A line extending from the mount into the third phalanx is significant of great intelligence and astuteness. ¶ 731. Line to third phalanx.

§ 5. *Signs on the Thumb.*

THUMB.

Signs are much rarer upon the thumb than upon fingers, but still they are sometimes found.

A subject who has several lines traced along the entire length of the phalanx of will, will make a faithful lover, having the gift of constancy and fidelity. ¶ 732. Several lines.

¶ 733. *Cross lines.* Cross lines upon the thumb denote riches.

¶ 734. *From ♀ to logic.* Lines extending from the mount on to the phalanx of logic, are a sure sign that the subject is much beloved.

¶ 735. *Star on logic.* A star on the phalanx of logic in a female hand is a sign of great riches.

¶ 736. *Ring round the joint.* A ring right round the joint which separates the phalanges of will and logic was held by the older cheiromants to be the sign of the scaffold.

SUB-SECTION VII.

THE TRIANGLE, THE QUADRANGLE, AND THE RASCETTE.

§ 1. *The Triangle.*

THE TRIANGLE.

THE triangle [called also the Triangle of Mars, from the fact that it is filled by the Plain of Mars] is the name given to the triangular space enclosed between the lines of life, head, and health. When [as is often the case, *vide* ¶ 629,] the line of health is not present in a hand, or so very badly traced as to be almost invisible, its place must be supplied by an imaginary line drawn from the base of the line of life to the end of the line of head, *or*, this side of the triangle may be formed of the line of Apollo.

¶ 737.
Its position and construction.

Though it must be considered as a whole, still each part of the triangle has its special signification ; thus, it is composed of the UPPER ANGLE formed by the

¶ 738.
Its composition

junction of the lines of life and head; the INNER ANGLE, formed by the junction of the lines of head and health; and the LOWER ANGLE, formed by the junction of the lines of health and life. [The lower angle may also be formed of the junction of the lines of health and of fortune.]

¶ 739. Neat and clean. If the triangle is well traced and neat, being composed of good even lines, [as in Plate XVII.,] it indicates good health, good luck, a long life, and a courageous disposition.

¶ 740. Large and well coloured. If it is large it denotes audacity, liberality of mind, generosity, and nobleness of soul; to have these significations it must be well and healthfully coloured, not livid, or approaching to deep red.

¶ 741. Small and curved. If it is small and formed of lines curving much inwards it betrays pettiness, cowardice, and avarice.

¶ 742. Its growth. Sometimes a triangle will form itself in a hand which began by being absolutely without it; this is a sign that the health, originally bad, has improved with advancing years.

¶ 743. Rough skin. If the skin inside the triangle is rough and hard, it is an indication of hardihood and strength of nerve.

¶ 744. Cross in the centre. A cross in the triangle denotes an extremely quarrelsome and contrary disposition. It betrays a state of mind best described by the American expression "cussedness." Many crosses in the triangle betoken continual bad luck.

¶ 745. Crescents. A crescent in the triangle, as at *f*, in Plate XVII., betrays an extremely capricious disposition, often indicating brutality and a love of bullying. If it is joined to the line of head, as at *g*, in the same figure, it is a prognostic of a violent death brought upon oneself by an imprudence or a want of calculation. Joined, however, similarly to the line of health, as at *h*, it is a sign of power and of success accompanied by excellent health.

A star in the triangle denotes riches, but riches obtained with much difficulty and worry. If the star is the termination of a worry line it indicates a sorrow, and if the worry line comes from a star in the Mount of Venus, it denotes that the sorrow has resulted from the death of a parent or of some near relation.

¶ 746. Star.

§§ 1. *The Upper Angle.*

UPPER ANGLE.

The upper angle [*i*, in Plate XVII.] should be neat, clearly traced, and well pointed; it indicates refinement and delicacy of mind.

¶ 747. Proper formation.

Blunt and short, it betrays a heavy, dull intellect, and a want of delicacy. *Very* blunt,—*i.e.*, placed under the Mount of Saturn,—it betrays a great danger of misery, and a tendency to avarice.

¶ 748. Bluntness.

The other extreme, however,—*i.e.*, *very* pointed,—is a sign of malignity, envy, and finesse.

¶ 749. *Very* pointed.

§§ 2. *The Inner Angle.*

INNER ANGLE.

The inner angle, [*k*, in Plate XVII.,] if clear and well marked, indicates long life and a quick intelligence.

¶ 750. Well marked.

Very sharp, it betrays a highly nervous temperament, and nearly always a mischievous disposition.

¶ 751. Very sharp.

Obtuse and confused, this angle denotes heaviness of intelligence, dulness of instinct, and, as a resulting consequence, obstinacy and inconstancy.

¶ 752. Blunt.

§§ 3. *The Lower Angle.*

LOWER ANGLE

The lower angle, [*l*, in Plate XVII.,] well defined, and just open, [as at *d*,] gives strong indications of good health and a good heart. If it is too sharp—in fact, if it is closed up—it denotes avarice and debility.

¶ 753. Proper formation.

If it is heavy and coarse, composed of many rays, or of a confusion of lines, it betrays a bad nature, with a strong tendency to rudeness and laziness.

¶ 754. Bluntness

Thus it will be seen that it is most important to observe, with reference to its component lines, the formation of the triangle and of its constituent angles.

QUADRANGLE.

§ 2. *The Quadrangle.*

¶ 755.
Its position and constitution.

The quadrangle [*vide* Map, Plate VII.] is the square space contained between the lines of heart and of head. It may be said to be bounded at its two ends by imaginary lines, drawn perpendicularly to the line of head from the crevice between the first and second fingers, and from the crevice between the third and fourth fingers.

¶ 756.
Proper aspect.

It should be fairly large and wide at the two ends, [but not too narrow in the centre,] clearly distinguishable, and of a smooth surface comparatively free from lines; under these aspects it indicates fidelity, loyalty, and an equable disposition.

¶ 757.
Narrow in centre.

Too narrow in the centre, it betrays malignity, injustice, and deceit, often accompanied by avarice. If it is much wider under the Mount of Mercury than under that of Saturn, it betrays a degeneration from generosity to avarice. Narrow under the Mount of Mercury, it denotes a more or less continual anxiety about reputation.

¶ 758.
Too wide.

Too large and wide throughout its extent, it signifies imprudence, or even folly; and this is so even when there are other signs denoting prudence in the hand.

¶ 759.
Much lined.

The quadrangle much filled up with little lines is a sign of a weak head.

¶ 760.
Badly traced.

If it is so badly traced as to be almost invisible as to its boundaries, it is a signification of misfortune, and of a malignant, mischievous character.

¶ 761.
"Croix Mystique."

It must be remembered [*vide* ¶ 695] that it is in the quadrangle that we search for the "Croix Mystique" [*q.v.*].

¶ 762.
Star.

A well-coloured and well-formed star is a great indication of truth and trustworthiness. Such a subject is pliable, and can easily be dealt with by fair means, [especially by women;] such subjects generally make very considerable fortunes by their own merit.

A line from the quadrangle to the Mount of Mercury betokens the patronage and protection of the great.

¶ 763. Line to ☿.

§ 3. *The Rascette and Restreintes.*

THE WRIST.

These are the names given to the wrist and bracelets of life [*vide* map]. According to some writers, the first or upper line only is called the rascette, the inferior ones being known as the restreintes; for my part, I prefer to name the entire region the rascette, and the lines traced across it the bracelets of life.

¶ 764. The rascette, its composition.

The bracelets of life are so called because each is said to be the indication of twenty-five to thirty years of life. I have found that in ninety-nine cases out of every hundred a bracelet of life gives about twenty-five to twenty-seven years of life, and even when the line of life is short a well-braceleted rascette will still ensure a long life to the subject.

¶ 765. The "bracelets of life."

Three lines clearly and neatly traced denote health, wealth, good luck, and a tranquil life. The clearer the lines the better is the general health of the subject.

¶ 766. Three clear lines.

If the first line is chained we find a laborious life, but good fortune resulting therefrom.

¶ 767. Chained.

If the lines are altogether badly formed it is a sign of extravagance.

¶ 768. Badly formed.

A cross in the centre of the rascette, as at *m*, in Plate XII., is a sign of a hard life, ending with good fortune and quietude.

¶ 769. Cross.

An angle in the rascette, as at *m*, in Plate XV., is a sign of inheritances and of honours in old age. To

¶ 770. Angle and cross

this will be added good health if a cross appear in this angle, as at *i*, in Plate XVI.

¶ 771.
Pointing upwards.
If the bracelets of life break into points converging towards the base of the line of Saturn, it is a sign of lying and vanity.

¶ 772.
Star.
A star in the centre of the rascette will mean inheritances in a lucky hand, but inchastity in a weak sensual hand.

¶ 773.
Voyage lines.
Lines from the rascette extending upon the Mount of the Moon signify voyages. A line right up to the Mount of Jupiter will signify a very long voyage indeed; in fact, the distance of the voyages may be told from the length of the lines. If the lines converge towards the Mount of Saturn, but do not join there, it is an indication that the subject will not return from the voyage. One of them, ending on the line of life, denotes probability of death upon the voyage. If the lines are absolutely parallel throughout their course the voyages will be profitable, but dangerous.

¶ 774.
Line to ☿.
A line from the rascette straight up to the Mount of Mercury is a prognostic of sudden and unexpected wealth.

¶ 775.
Line to ☉.
A similar line going to the Mount of Apollo is a mark of the favour and protection of some great person.

¶ 776.
Line to health through ☽.
A line from the rascette near the percussion of the hand, passing through the Mount of the Moon to join the line of the liver or health, is a sign of sorrow and adversity, especially if the line be unequal and poorly traced.

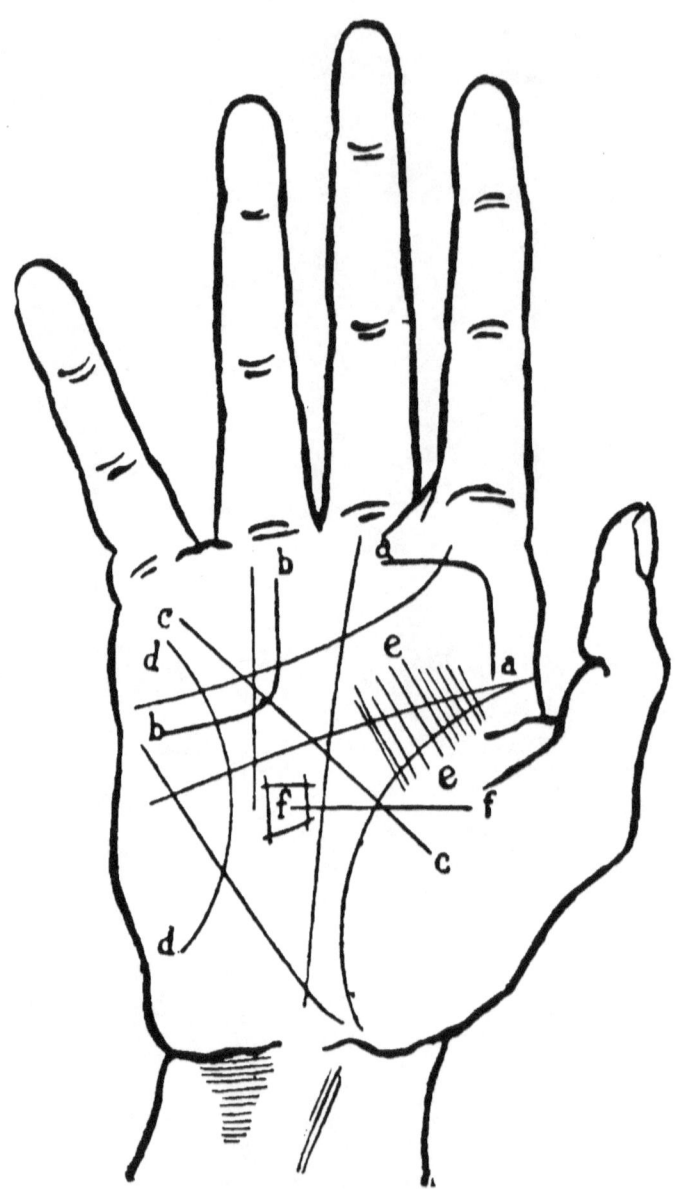

PLATE XVIII.—CHANCE LINES.

SUB-SECTION VIII.

CHANCE LINES.

CHANCE LINES

¶ 777.
Definition of chance lines.

WE have now arrived at a point from which having carefully discussed all the regular details of cheirosophy, it is necessary that we should turn to the consideration of certain lines which appear from time to time in the hand, and which, having special significations of their own, cannot be taken account of whilst going categorically through the indications of the principal lines, and of the various combinations of them. Their number is, of course, practically unlimited, for they form themselves according to the characters and lives of individual subjects. The student will find after a time that as the groundwork of cheiromancy impresses itself upon him, he will be able at once to read the indications of any line which may be shown to him, though he may never have seen one like it before. The following instances, therefore, are not given as being in any way a complete list of the "chance lines," but are subjoined as a kind of guide for the student, to enable him to decipher these "eclectic indications" whensoever and wheresoever he may find them. The following instances are, for the most part, illustrated in Plates XVIII., XIX., and XX., so that there will be no difficulty in remembering their exact positions. In these figures the principal lines are *drawn*, but only the chance lines are lettered and referred to.

¶ 778.
From life to ♃ and ♄.

A line starting from the commencement of the line of life, going to the Mount of Jupiter, and then turning on to the Mount of Saturn, as at *a a*, in Plate XVIII., denotes a disposition to fashionable fanaticism. If such a subject is religious *at all* it will be, that he is actuated mainly by a desire to become eminent in that particular line.

¶ 779.
From ♂ under heart to ☉.

A line starting from the Mount of Mars, running under the line of heart, and turning up to the Mount of Apollo, as at *b b*, in Plate XVIII., indicates a determination to attain celebrity so deeply rooted, that the subject whose hand bears this line will attain that celebrity by *any* means.

¶ 780.
From ♀ to ☿.

A line barring the whole hand from the Mount of Venus to that of Mercury denotes cleverness and intelligence, arising from an affair of the heart, or from the promptings of passion.

¶ 781.
Worry lines.

We have in another place discussed worry lines, [*vide* ¶ 533,] which are, after all, a species of chance line; any worry line which starts from a star on the Mount of Venus denotes that some one very dearly beloved has died.

¶ 782.
From ♀ to ♂.
Star.

Two worry lines, extending parallel from the Mount of Venus to that of Mars, denote the pursuit of two love affairs at the same time, and a star joined to these lines denotes that the pursuit has ended in disaster.

¶ 783.
Curved from ☿ to ☽.

A curved line extending from the Mount of Mercury to that of the Moon, [as at *d d*, in Plate XVIII.,] is a signification of presentiments and occult powers. Such a subject, if his line of head decline upon the Mount of the Moon, will have great powers as a medium.

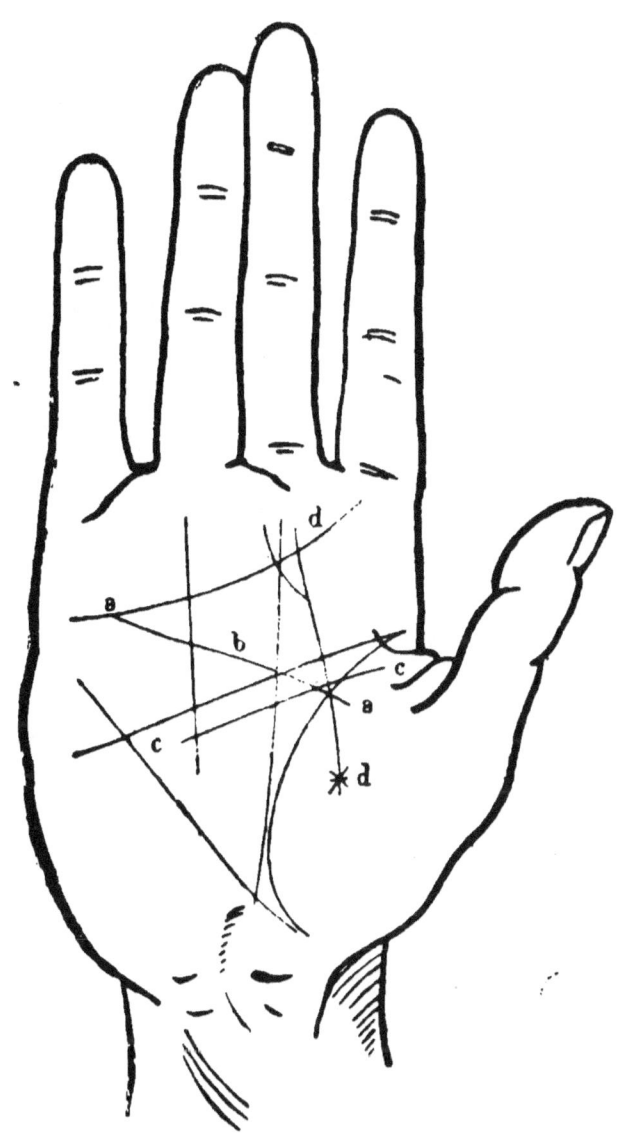

PLATE XIX.—CHANCE LINES.

CHANCE LINES.

If, with a chained line of heart, a line from the Mount of Venus touch it underneath the Mount of Mercury, [as at *a a*, in Plate XIX.,] it is a sign that the whole life has been disturbed and worried by a woman [or *vice versâ* in a female hand]. A black point on this line [as at *b*, in Plate XIX.] signifies widowhood or widowerhood.

¶ 784. Chained heart and line from ♀ to ☿. Point.

A line from the Mount of Venus cutting the line of Apollo, [as at *c c*, in Plate XIX.,] denotes a misfortune at the time indicated by the point at which the line cuts through the line of life. If it cuts through in early life, the misfortune was connected with the parents of the subject.

¶ 785. From ♀ to line of ☉.

Quantities of little rays across the line of life into the quadrangle, [as at *e e*, in Plate XVIII.,] accompanied by short nails, are a certain sign of quantities of little worries, estrangements of friends, etc., occasioned by the spirit of argument and criticism, and the love of teasing which the subject has, by reason of his short nails.

¶ 786. Rays across the line of life.

A line extending from a star on the Mount of Venus to a fork under the finger of Saturn [as at *d d*, in Plate XIX.] betrays an unhappy marriage.

¶ 787. From ♀ to ♄.

A line starting from the Mount of Venus, and ending in a square in the palm of the hand, [any part,] as at *f f*, in Plate XVIII., is significant of a narrow escape from marriage with a scoundrel, or with an extremely wicked woman.

¶ 788. From ♀ into palm.

A long island, extending from the Mount of Venus to that of Saturn, with a similar island in the line of fortune, both at the points representing the same age, [*vide* Plate XI.,] as at *a* and *b* in Plate XX., indicate seduction.

¶ 789. Island from ♀ to ♄.

A line going from a star on the Mount of Venus to the Plain of Mars, and then turning up to the Mount of Apollo, where it meets a single ray, [as at *c c*, in Plate XX.,] foretells a great inheritance from the death of a near relation.

¶ 790. From ♀ to ♂ and ☉.

¶ 791.
Many little lines on the percussion.

A *quantity* of little lines on the percussion, at the side of the Mount of Mercury, [as at *d*, in Plate XX.,] indicate levity and inconstancy, [*vide* also ¶ 469,] especially if the Mounts of Venus and of the Moon are highly developed.

¶ 792.
Method of interpretation.

These few instances will, I am sure, be sufficient to explain the method of interpreting chance lines. It will be observed that they are read carefully with reference to the mounts and lines which they cross throughout their course, and according to the signs which meet and interrupt them.

The student has now traversed the entire field of Cheirosophy. It only remains for me now to give my readers a sub-section containing a few illustrative types, before closing this manual with a few remarks on the method of proceeding in making a cheiromantic examination of a subject.

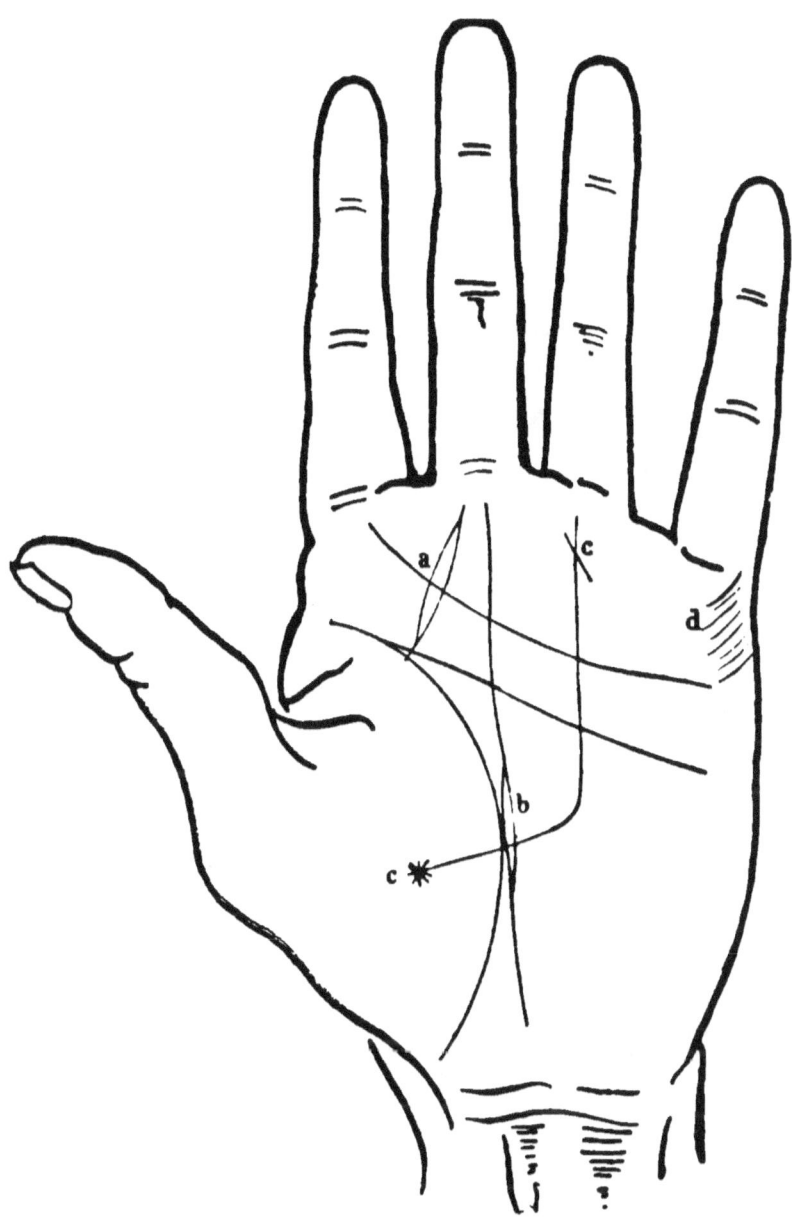

PLATE XX.—CHANCE LINES.

SUB-SECTION IX.

A FEW ILLUSTRATIVE TYPES.

I PROPOSE in this section to describe a few types of character and of profession; that is to say, I propose to set forth the collected signs and formations which indicate certain conditions of mind, with the probable effects of those conditions upon the subject, as regards his choice of a profession, or his walk in life.

For instance: take a hand which betrays a murderous or homicidal tendency; in this hand you will find the general complexion to be very red, or very livid; if the former, the tendency to murder arises from fury and momentary fits of anger; if the latter,

ILLUSTRATIVE
TYPES.

¶ 793.
Collective
indications.

¶ 794.
Homicide.

the whole instinct of the subject is evil. The first phalanx of the finger of Mercury will be heavily lined, and at the base of the line of life will [probably] be found a sister line. The line of head will be deeply traced and thick, having probably a circle upon it, and being generally joined to the line of heart, and separate from the line of life. The nails short, the line of life thick at the commencement, and spotted with red, and the line of head twisted across the hand. All these signs collected in a hand are an infallible indication of a murderous instinct.

¶ 795.
Theft.

Take another example : in this hand we find the line of head twisted and very red, a grille is placed upon the Mount of Mercury, and the whole hand is dry and thin, having the joints developed on the fingers. From the third phalanx of the little finger sundry small lines go on to the mount, which latter is also scarred with a deep strong ray. This is the hand of a thief, and the impulse of theft will be found to be almost [if not quite] insurmountable.

¶ 796.
Falsehood.

Falsehood,—*i.e.*, a general tendency to deceit—is always very clearly marked in the hand, and is marked by a number of different signs, any one of which by itself is a sufficient indication of a strong tendency in that direction. These are : a high Mount of the Moon, upon which the line of head is forked, and on which are found small red points ; the thumb is short, and on the inner surfaces of the phalanges of the fingers there appears a kind of hollowing out or sinking in of the flesh. The line of head is generally separated from that of life by a space which is filled with a number of confused lines.

¶ 797.
Application of Cheirology.

In conversing with a subject in whose hands you have seen all, or any, of these signs, bear in mind what we have said under the heading of Cheirology [SS. I., § 11].

A FEW ILLUSTRATIVE TYPES. 303

Another very characteristic hand is the voluptuous, or pleasure-loving hand. The fingers are smooth and pointed, having the third, or lower phalanges, swollen; the whole hand is plump and white, the palm strong, and the thumb short, giving it sensitiveness. The Mount of Venus is high. Such subjects are impressionable, and liable to fall into grave errors; they are sensual, vain, and egoists, always actuated by motives of pleasure. Women who have these hands are always dangerous, for they are subtle and unscrupulous in their pursuit of enjoyment, and often exercise a most fatal influence upon men into whose lives they come.

¶ 798. Sensuality.

Adrien Desbarrolles, in his later and larger work on the science, [*vide* Note [119], p. 233] devotes a considerable space to the indications of various professions. It would be beyond the scope of a work like the present one to go into the matter as fully as he does at page 350 of that volume, but a short *resumé* of his leading principles may not be out of place in a chapter on illustrative types.

¶ 799. The professions according to Desbarrolles.

Of an artist, the sign is of course primarily the artistic hand, [*vide* p. 148,] but our author goes further. He discusses the various modifications which betoken different classes of painting; thus:—the flower painter will have the Mount of Venus high with long fingers, and a large thumb; [colour, detail, and perseverance;] the painter of still life will have rather squared fingers and the Mount of Mercury; [exactitude and science;] the painter of battle pieces will have the Mount of Mars developed, indicating the natural taste of the subject. He points out the fact that painters with squared fingers always paint what they can actually see rather than what they merely imagine.

¶ 800. Artist's hand.

In a doctor's hand we shall find the Mount of Mercury rayed with the line of Apollo clearly traced.

¶ 801. Doctor's hand.

The doctor whose hands bear the Mount of the Moon well developed will always be inclined to discoveries and eclecticism, and the doctor with hard hands and very much spatulated fingers will have a natural *penchant* for veterinary surgery.

¶ 802. Astronomy. The astronomer has the Mounts of the Moon, of Mercury, and of Saturn well developed, with long knotty fingers to add calculation to his imagination and his science.

¶ 803. Horticulture. The horticulturist has a hand in which we find the Mounts of Venus and of the Moon high; with spatulate fingers to give him energy, and long fingers to give him detail.

¶ 804. Architecture. Square fingers, with a good line of Apollo and a good line of Jupiter, denote an architect.

¶ 805. Sculpture. Sculpture betrays itself by a scarcity of lines, the Mounts of Venus, of Mars, and of the Moon high in the hand, which has a strong tendency to thickness and hardness.

¶ 806. Literature. Literary men have always the Mounts of Jupiter and of the Moon developed; the latter particularly, if the taste lies in the direction of poetry. Literature gives, as a rule, soft spatulate or square hands, with the joints [especially that of matter (the second)] slightly developed. Literary critics have always short nails and high Mounts of Mercury.

¶ 807. Music. Among musicians [*vide* ¶ 303] execution is the domain of subjects whose fingers are spatulate, and whose Mount of Saturn is high, whose nails are short, and whose joints are developed, with the Mount of the Moon prominent, long thumbs, the Line of Apollo, and [as a rule] the Girdle of Venus. Melody generally gives smooth fingers with mixed tips, the prevailing mount being that of Venus.

¶ 808. Drama. The actor has fingers which are either spatulate or square, the Mount of Venus developed, and the line of head forked. The line of heart turns up slightly towards the Mount of Mercury, and, as a rule, a line runs from the Mount of Mars to that of Apollo.

I have selected the above illustrative types from those given by M. Desbarrolles, as being those which, by repeated and careful examination, I have found to be, with extremely few exceptions, completely correct. Their explanations are easily found, [*vide* ¶ 108,] and the student will, in a very short time, be able, immediately on seeing a hand, to tell the subject what is his profession.

SUB-SECTION X.

MODUS OPERANDI.

MODUS OPERANDI.

¶ 809. *Uselessness of the science without knowledge of how to practise it.* MANY years ago I bought at a marine store a second-hand sextant. I was not going a long journey, and I had absolutely no need of the sextant, but I bought it because it was a beautifully finished instrument, because there was something strange, incomprehensible, mysterious, and therefore fascinating, about it, and because it was very cheap. When I had got it I did not know what to do with it; I could not use it, for I knew not how, and following the ordinary course of things, it was put away to get rusty and impracticable, without ever having been of the slightest use to me. I mention this apparently irrelevant circumstance because whenever I see a work on cheirosophy in the possession of any one I always think of my sextant, and wonder whether they, too, having taken up the science of cheirosophy because it is strange, apparently incomprehensible, mysterious [to them,] and therefore fascinating, have any idea of how to put their knowledge into operation, or whether, after playing inquisitively with the science for a time, they will let it lie by and become rusty and useless. It is urged by these considerations that I have decided to write this very important sub-section, so as not to load my readers with a quantity of knowledge, with a complicated instrument, that they cannot make use of, and derive a practical benefit from.

¶ 810. Condition of the hands.

Much has been said in works on cheiromancy on the condition of the subject at the time of the examination, his mental and physical state, and so on, but I think that all these things are, to a very great extent, immaterial. The only things to be borne in mind are [*selon moi*] that the hands should not be too hot or too cold, and that they should not have just been pulled out of a tight glove, and, above all things, that there should be a good light. The hand should be held in an oblique position as regards the light, so as to throw the lines and formations into relief. With this object in view, also, the fingers should be slightly bent, so as to contract the palm and accentuate the lines, for it must be observed that the hands fold upon the lines, though the lines are not formed by the folding. If it is quite convenient, the morning is the best time to examine a hand, but it is practically immaterial if the cheirosophist has had any experience.

¶ 811. Mode of procedure.

Lastly, in reading a hand, to whomsoever it belong, you must never hesitate to take it in your own hands and hold it firmly. These short preliminaries being attended to, you will commence your examination. It is far better to examine the whole hand carefully and silently till its indications are quite clear in your own mind, and then to speak promptly and boldly, than to decipher the indications slowly one after another, reading one tentatively, with a view to ascertaining its correctness, before going on to another.

¶ 812. Simplicity of the science.

The great thing that I desire to impress upon the minds of my readers is the simplicity of the science. Adrien Desbarrolles, in his advanced work on the science, says: " That which prevents beginners from succeeding immediately in cheiromancy is that they find it too simple, and think it necessary to go beyond it to arrive at something more pretentious, more confused, more difficult, and more impossible to understand. They do not *want* an easily understood science. For many people, a science which is simple, is not a science at all; they strive and strive, racking their brains in search of a truth which is at their very hands, and which they can find nowhere else."

¶ 813.
Order of the examination.
Cheirognomy:

Having taken a hand in yours, first you must examine the line of life, to see what effects health and the great events of life have had upon the condition of the subject. Next look at the phalanx of will, and see how far it is controlled or influenced by the phalanx of logic. Then you will note the tips of the fingers, seeing also whether they are smooth or whether they have the joints developed, and whether any particular phalanx or set of phalanges is or are longer or more fully developed than the others; this will tell you whether the subject is governed by intuition, by reason, or by material instinct. Then notice whether the fingers are long or short. At first you can hardly tell whether they are long or short, but after a little time you will be able to judge at once of length or shortness by comparison with the other hands you have seen; the same remarks apply to the thumb.

¶ 814.
Order of examination.
Cheiromancy.

You have already noticed whether the hands are soft or hard, now you will turn your attention to the palm, to see what mount or mounts govern the instincts, and how those mounts are governed in turn by primary or secondary lines. Then go back to the line of life, and examine the line of fortune, noting whether the latter is broken, and if so, search on the mounts for signs to teach you the cause and interpretation of the break. Then examine carefully the lines of head and heart, and the secondary lines with the signs which may modify their indications. Be careful not to predict a future event from a sign which is evidently that of a past one: a sign which, though visible, is effaced or quasi-effaced, is that of a past event; a sign which is clear and *well coloured* is that of a present circumstance; and a sign which is only just visible, as it were, beneath the surface of the skin, is that of a future event.

<small>Past, present and future.</small>

Whenever you see a star, a cross, or any other sign in an apparently inexplicable position, you must search the principal lines and the mounts for an explanation. The explanation will often be found in a mark on the line of fortune or in a worry line [*vide* ¶ 533]. At the same time look at the Mount of Jupiter, for this will often, by being good, counteract the evil indications of a sign, and at the Mount of Mars to see whether the subject has that resignation which will give him calm, and even happiness, through whatever circumstances may assail his life.

<small>¶ 815. Uncommon signs.</small>

When you have examined everything, strike a balance, as it were, noting what signs are contradicted or counteracted by others, and what is, in fact, the whole indication of the hand. Speak boldly, and never mind offending people by what you tell them; what you tell them is *the truth*, and they need not have let you know it. I always warn people that what I shall tell them will be the actual truth, and not a string of complimentary platitudes; and I always ask people not to show me their hands if they have anything to conceal. If, after this, they still persist in having their hands read, I say boldly whatever I see there, without caring about the feelings of the subject.[125]

<small>¶ 816. Exposition of the indications</small>

[125] "Physicians are, some of them, so pleasing and conformable to the humour of the patient, as they press not the true cure of the disease; and some others are so regular in proceeding according to art for the disease, as they respect not sufficiently the condition of the patient. Take one of a middle temper, or, if it may not be found in one man, combine two of either sort ṭ

¶ 817.
Envoi.

With an account [by way of illustration of the above maxims] of an interesting experience of mine now some years old, I close this sub-section, and with this sub-section this manual. I have had but one object in view during its composition—clearness; I can only hope that I have been successful in attaining that object, and if my introductory argument has had any power to commend this science to the consideration of unbiassed and discriminating minds, I shall feel that I have not wasted the labours of the years that are past.

Cheirosophic Experience.

A few years ago I had left my papers and goose-quills on a magnificent summer's afternoon, and had betaken myself to a garden-party in one of our prettiest suburbs. As has often been the case, my arrival was the signal for a chorus, "Here's the Cheiromant, let's get our fortunes told." In vain I remonstrated that I was out for a holiday, in vain I pleaded ill-health, ill-temper, and ill-manners; the torrent of silvery persuasion still flowed on, till at last I said :—

"Listen to me; if it will amuse you, I will read *one* pair of hands for you, but they must be those of a complete stranger, and no one must ask me to repeat the experiment." Half a-dozen pairs of hands were put forward, and from among them I chose those belonging to a childish face and a mass of sunny hair, which I had certainly never seen before. I said to her—"If you like, I'll read your hands for the amusement of all these people, *but*, before I begin, if there

and forget not to call as well the best acquainted with your body as the best reputed of for his faculty."—FRANCIS BACON, "Of Regimen of Health," 1625.—'Καὶ ταῦτ' οὐχ ὑπ' ἀπέχθωμαι τίσιν ὑμῶν, τὴν ἄλλως προῃρῆμαι λέγειν. οὐ γὰρ οὕτως ἄφρων οὐδ ἀτυχής εἰμι, ἐγώ, ὥστ' ἀπεχθάνεσθαι βούλεσθαι μηδὲν ὠφελεῖν νομίζων. ἀλλὰ δικαίου, πολίτου κρίνω τὴν τῶν πραγμάτων σωτηρίαν ἀντὶ τῆς ἐν τῷ λέγειν χάριτος αἱρεῖσθαι.—DEMOSTHENES, ΟΛΥΝΘΙΑΚΟΣ. Γ" (21).

is anything in your life that you have the least objection to the whole world knowing, say so at once, and I'll read some one else's."

"Oh dear no!" replied she; "my life has been utterly uneventful, go on." I looked at her hands, then at her,—still the laughing, childish face, and the calm, untroubled eyes,—and said :—

"How dare you tempt Providence like this out of pure bravado? *You know* perfectly well that there are events in your life which you *don't* want every one to know, and yet disbelieving [at present] in a science of which, [knowing nothing of it,] you are not in a position to give an opinion, you hold out your life's history for the amusement of a garden-party crowd. If you still insist, I will tell you your life *here* and *now*, but I should suggest that we should take a turn round the lawn, and *then* you will come back and tell these people that everything I have told you is absolutely correct."

She thought for a moment, and said: "There is nothing I am ashamed of; but in case you are making some horrible mistake, I'll hear my hands read in private." So we walked round the lawn.

"Well," said I, "you look about nineteen, and as if you had never had a trouble in your life, but you have had the most terrible time of it I have ever seen written on a hand so young as yours. You have been married twice, and this, of all things, strikes me in your hand, that you married your second husband when your first husband was alive. Your first marriage was an affair of pique, an impulse of your foolish head, and was a miserable one; your second was an affair of heart, a love-match, but it was extremely bad for you from a commercial, material point of view. Even now, whilst you walk here with a smile on your lips and a racquet in your hand, you are undergoing some mental agony: let me congratulate you on being the most astounding—actress, shall I say?—that I have ever come across."

She was silent for a moment, and then said: "What I am going to tell you no one but my maid, who is in Chicago now, has ever known, and I tell it you as a reward for speaking so boldly in the face of the magnificent lie I told you just now. I am an American, and came here with some people to-day, and don't know a soul in the place; I am twenty-three [though I don't look it]. At eighteen I quarrelled with my people, and in a fit of rage married, simply to get rid of them. My husband turned out a scoundrel and knocked me about, to speak plainly, and after a year we were divorced. When I was twenty I fell in love for the first time, and married a man whom I simply worshipped. We were as happy as possible, but after a few months he was struck with a fever that gradually wasted him away, and he died two years ago, leaving me simply a pauper, for during his illness his business in Chicago left him. I came over here with some friends. What you say about my present state of mind is quite true, for I saw my first husband yesterday at the Academy, and have been in a state of terror ever since. *Now*, if you please, we will go back and lie to the other people about what you have been telling me."

The lady left for Yokohama a few months later, and sent me her permission to publish this incident.

FINIS.

INDEX I.

OF MATTERS REFERRING TO THE SCIENCE OF CHEIROSOPHY, CONTAINED IN THE INTRODUCTION AND THE NOTES THROUGHOUT THE VOLUME.

The numbers refer to pages. A number thus signalized—"32 [18]"—refers to *Note* 18 on *page* 32.

Names of Authors, whose works are referred to with their titles, etc., are printed in *italics*.

For reference Index of Cheirosophic Indications vide *Index II*.

A.

Abercrombie, J., 51.
Abuse of the science, 7 [6].
Adaptation of the hands to the wants of man, 20.
Æther, 63.
Alteration of hands, 73.
Anatomy, 38.
Anaxagoras on the hand, 23.
Ancient and modern hands, 28.
Anthropology of the subject, 27.
Apollo Belvidere, The, 28 [12].
Aristotle on the hand, 23, 24, 29, 36, 44, 47, 59, 60, 70, 82, 98, 101, 131.
Aristotle on Cheiromancy proper, 60.

D'Arpentigny, S., 18 [1], 71, 74, 84, 117.
Arguments in favour of Cheirosophy, 86.
Arteries of the hand, 41, 43.
Astral lines, 67.
Astrologic Cheiromancy, 63.
Astrology, 65.
Astronomical parallel, 78.
Authorities for phenomena, 87.

B.

Bacon, F., 90, 92, 218 [117], 309 [1,5].
de Balzac, 62, 74, 75, 82, 88.
Balzac on Cheiromancy, 60, 66.
Beamish, R., 27.

Bell, Sir C., 20 [2].
Bell on the hand, 20, 24, 25, 41, 51, 52, 71.
Bernstein, J., 45.
Bible Cheiromancy, 53, 58.
Bibliography, 60.
Births of the sciences, 88.
Biting the thumb, 34.
Blessings, 32.
Bones, Identification of, 71.
Bones of the hand, 37.
Brain, Connection with the, 49.
Browne, Sir T., 134.
Bulwer, J., 129 [103].
Burton, R., 131 [106].

C.

Cæsar, 32.
Cardan, 83, 132 [109], 134.

INDEX.

Cause and effect, 65, 66, 68, 81.
Challenges with the thumb, 34.
Charlatanry, 76.
Charms, 33.
Cheirology, 129.
Chinese divination by the thumb, 69.
Cicero, 26, 189 [112].
Classic authorities, 59, 186 [112].
Cocles, B., 83, 84 [79].
Commentators of the Bible, 53-5.
Comparative anatomy, 38.
Construction of the hand, 28, 36.
Consumptive finger-nails, 52.
Contracts, Use of the thumb in, 34.
Criminal without hands, 520 [3].
Currents of impression, 80.
Customs in the East, 29, 31.
Cutting off the hand, 31.
Cuvier, G., 47.

D.

Daubenay, 64.
Degeneration of the science, 76.
Demosthenes, 82, 97, 183, 310
Desbarrolles, A., 17, 50, 67, 78, 83.
Differentiation of nerves, 89.
Disease, Indications of, 51.
Distinctness of the hand, 36.
Dodd's execution, 34.
Dryden, 59.
Dumas fils, A., 68.

E.

Earth and the Planets, 63.
Ebule, Evans A., 73-79.
Electricity simile, 50, 51.
Epicurus, 44.
Esdaile's experiments, 92.
Esdaile, Dr., 92 [69].
Events, Laws of successive, 65.
Evil eye, The, 33.
Explanations unnecessary, 89.

F.

Ferrand, J., 77.
First finger, The, 32.
Five, The number, 30.
"Folding" argument, The, 71.
Folding of hands in prayer, 29.
Foot, The, 37.
Force denoted by word hands, 29.
Force, Regulation of, 25.
Foretelling events by Cheirosophy, 81, 84.
Formation of tissues, 40.
Fresnel, A., 64.
Future lines, 81, 82.

G.

Galen on the hand, 20, 21, 24, 26, 40.
Gall, F. J., 85.
Gaule, J., 110.
Gellius, J., 42.
Genuineness of the science, 91.
Georget, 52.
Germs of disease, 81.
Giving the hand, 29.
Gladiatorial signs, 34.
Greek use of the word χείρ, 28.

H.

Halfway sceptics, 73.
Hand charms, 33.
"Hand of Glory," 35.
"Hand of God," The, 28 [13].
Hand superstitions, 34.
Hand shaking, 19.
Hartlieb, J., 60.
Herder, 51.
Hereditary Cheirosophy, 75.
Heron-Allen, E., 77 [71], 91 [98], 122 [102].
Holiday, B., 116 [99].
Holy Scriptures and Cheiromancy, 53.
Homer, 25, 30.
Homogeneity, 70.
Honesty of the science, 91.
Human Infant, The, 51.
Humphry, G. M., 36, 38.
Dr. Hunter, 51.
Huygens, Z. van, 63.

I.

Ignorant incredulity, 92.
Impatience in argument, 76.
Importance of the study, 17.
Incapacity for comprehension, 88.
Indications in youth, 79.
Indications of bodily aspects, 84.
Indications of disease, 51, 52.
Indication of race, 27.
Infant hands, 72.
Infant hand and thumb, 40.
Infant, The human, 51.
Infinite differences of hands, 69.
Ingoldsby legends, 35.
Interpretations of appearance, 71.
Investigation of phenomena, 86.
Investigation of the science, 92.
Identification of bones, 71.

J.

Jettatura, The, 33.
Juvenal, 59.

K.

Kirchmann, J., 42.
Kidd, Dr., 21 [4], 47.
Kissing hands, 29, 30.
Kollmann, A., 48.

L.

Lavater, J. C., 69 [66], 74 [69], 84, 85 [80].
Lemnius, Levinus, 42.
Light, Theories of, 63.
"Line-roots," 81.
Lines caused by folding the hands, 72.
Lines in the hand, 49, 80.
Loss of the hands, 20.
Love of the Marvellous, 94
Lucretius, 26.

M.

McDougall, F. T., 51.
Mano Pantea, 33.
Manufactures by the hand, 22.
March of enlightenment, 91.

INDEX.

Marriage service, The hand in the, 32.
Marvellous, Love of the, 94.
Massey, C. C., 87.
"Materialistic" opposition, 90.
Melton, J., 135.
Modifications of types, 73.
Montaigne, 61.
Motor and sensory nerves, 52.
Muller, 51.
Multiplied powers of the hand, 61.
Muscles of the hand, 38, 39.
Muscular sense, The, 25.

N.

Narrow-minded sceptics, 93.
Natal influences of the Planets, 67.
Nerves and electricity, 50.
Nerves of the hand, 44.
Newton, I., 63.
Noel, Roden, 86.
Non-fatalism of the science, 81.
No two pair of hands are alike, 69.
Number five, The, 30.

O.

Oaths, 31.
Obstacles to the study, 93.
Obstinate incredulity, 92.
Offensive and defensive functions of the hand, 25.
Organ of mind, The, 24.
Oriental customs, 29, 31.
Osteology, 37.
Owen, Professor Sir R., 23, 36, 44, 83, 207 [113].
Owen on the foot, 37.

P.

Pacinian bodies, 49.
Painful side of the science, 78.
Peace, Signs of, 29, 33.
Perfection of the hand, 26.
De Peruchio, 116.
Petit Albert, 35.
Phrenology and Physiognomy compared with Cheirosophy, 73.
Phrenology, 85.
Physiognomy, 85.

Physiology of the hand, 36.
Planets, Influence on the earth of the, 63.
Playfair, L., 88 [64a].
Pliny, 30.
Poltroons, 34.
Powers of the hand, 61.
Prediction, 81, 82.
Progress of the science, 91.
Punishment of hand-amputation, 31.
Purkinye, J. E., 49.

Q.

Quintilian, O. F., 61.

R.

Racial hand, 27.
Radial artery, 43.
Raising the hands, 33.
Registration of cases, 87.
Regulation of force, 25.
Religion, 87.
Religious opposition, 93.
Religious symbols, 32.
Revealed sciences, 87.
Ridicule, 69, 75.
Ring finger, The, 41.

S.

Sabine, Experiments of Col., 65.
Sceptics, 73, 76, 92.
Scientific opposition, 93.
Scipio Africanus, 32.
Scott, Sir W., 45.
Second finger, The, 32.
Sensation of pain, The, 46.
Sense, The muscular, 25.
Sense of touch, 45.
Senses of the skin, 48.
Sensory and motor nerves, 52.
Sequence of events, 65.
Shakespeare and the word "hand," 29.
Shaking hands, 19.
Skin, The, 48.
Small beginnings in science, 87.
Spencer, H., 66, 76, 88, 89.
Spurzheim, Dr., 85.
Stewart, Dugald, 65, 90.
Study of Cheirosophy, The, 17.
Superiority of Cheirosophy, 75.

Superstitions of the hand, 34, 41.
Swearing by the hand, 31.
Symbolical expressions, 28.
Symbols, 28, 32.

T.

Tactile Corpuscles, 49.
Temperature, 52.
Third finger, The, 32, 41, 189 [112].
Thumb, The, 34, 39, 40.
Touch and pain, 46.
Touch a universal sense, 47.
Touch, Sense of, 45, 53.
Traditional Cheiromancy, 83.
Tricasso, 84 [79], 134.
Trickery in psychic phenomena, 86.
Trinity, The, 32.
Turkish rosary, 31.

U.

Ulnar artery, 43.
Unnecessary explanations, 89.
Use of an organ shown by its aspect, 70.
Use of the hands, 19, 22, 61.

V.

Value of Cheirosophy, 78, 79.
Vascularity, 53.
Venous system of the hand, 41, 43, 53.
Voting with the hand, 31.

W.

Warnings against Cheirosophy, 78.
Wickedness of the science, 76.

X.

Xenophon, 31, 32 [18], 34.

Y.

Young, Dr., 64.
Youth, Value of Cheirosophy in, 79.

Z.

Zauberflote, 77.

INDEX II.

OF THE LEADING INDICATIONS GIVEN IN SECTIONS I. AND II.

The numbers refer to the paragraphs. Thus:—Abstraction ¶ 254.

A.

Abstraction, 254.
Acrobats, 339.
Action, 156, 236.
Administrative talent, 119, 133, 253, 342, 706.
Adultery, 680.
Adventurous, 233.
Æsthetic, 166.
Affectation, 140.
Affection, 211, 259, 551.
After-dinner speakers, 241.
Age, 525, 604.
Aggressiveness, 449.
Ambition, 222, 223, 538, 582.
Analysis, 121.
Ancien Noblesse, 278.
Angles, 738.
Apathy, 346.

Apollo, Mnt. of, 450; Artistic instinct, 450; Boastful, 454; Discoverers, 451; Failure, 456; Fortune, 450; Materialism, 455.
Apoplexy, 565.
Architect, 804.
Argument, 589.
Arithmeticians, 275.
Artist, 137, 328, 800.
Artistic tendency, 123, 129, 163, 233, 238, 294, 295, 346, 621.
Asceticism, 441.
Assassination, 662.
Asthma, 521.
Astronomy, 802.
Astuteness, 731.
Athletics, 205.
Avarice, 152, 556, 575.

B.

Bad end, 617.
Barristers, 461.
Beautiful, 291.
Beloved, 734.
Bigotry, 348, 696.
Biliousness, 488.
Blindness, 628, 676.
Bohemianism, 291.
Boldness, 141.
Brutality, 113, 633.
Builders, 143.
Bullying, 745.
Business capacity, 158.

C.

Callousness, 228.
Capillaries, 427.
Caprice, 637.
Carelessness, 544, 582.

INDEX. 317

Caution, 580.
Celebrity, 542, 727, 779.
Charlatanry, 429.
Chastity, 513.
Children, No. of, 469.
Clairvoyance, 462.
Classical scholarship, 438.
Cleverness, 724, 780.
Colonists, 274.
Commercial talent, 237, 245, 600.
Common sense, strong, 188.
Communists, 277.
Comprehension of detail, 142.
Conical, 155, 163.
Conjugal infidelity, 613.
Conjugal misery, 608.
Constancy, 294, 578.
Coolness, 578.
Cowardice, 140, 177, 741.
Criticism, 253.
Cruelty, 149, 175, 718.
Cunning, 240, 244, 418, 641.
Curiosity, 151.

D.
Dancers, 339.
Danger, 567.
Deafness, 592.
Death, 470, 525, 529, 536, 557, 567, 572, 576, 579, 585, 645, 664, 675, 710, 717, 745.
Debility, 713, 753.
Deceit, 140, 757.
Delicacy of mind, 212, 255, 747.
Designers, 143.
Desire of life, 546.
Dictator, 298.
Dignity, 248.
Diplomacy, 350, 685.
Discontent, 338.
Disciplinarian, 297.
Dishonesty, 141, 590, 659, 682, 690.
Dissatisfaction, 421.
Dissimulation, 660.
Distrustful, 137.
Doctors, 439, 801.
Domestic troubles, 582.
Drowning, 505.
Dull intellect, 748, 752.

E.
Early rising, 205.

Eclecticism, 320, 777.
Effeminacy, 181.
Egoism, 122, 221, 273, 487, 586, 701.
Egyptian papyri, 143.
Eloquence, 729.
English character, 281.
Enthusiasm, 563.
Envy, 713.
Epilepsy, 587.
Error, 226.
Evenly - balanced mind, 113.
Excitement, 291, 702.
Extravagance, 196, 725, 768.

F.
Failure, 114, 456.
Faithful lover, 732.
False conceptions, 125.
Falsehood, 713, 796.
Fanaticism, 330, 437, 562, 688, 778.
Fancy, 123.
Fatalist, 433.
Fault finding, 283.
Feebleness, 112, 423, 572.
Ferocity, 257.
Fidelity, 756.
Fingers, 117, 123, 124, 126, 127, 131, 133, 134, 157.
First impressions, 123.
Flattery, 429.
Flirt, 554.
Foolhardiness, 544.
Foolishness, 199.
Freethought, 437.
Friendship of great men, 714.
Frivolity, 234.

G.
Gallantry, 507.
Gambler, 233.
Generosity, 292, 740.
Goodness, 594, 724.
Good digestion, 630.
Good fortune, 217, 439, 588, 606, 698, 715, 769.
Good health, 523, 630, 739, 753.
Good luck, 424, 562, 598, 642, 656, 662, 739.
Good sense, 242, 311, 571, 667.
Good talker, 305.
Gout, 498.
Grave, 414.

H.
Hands, 145, 150, 175, 176, 177, 178, 179, 180, 201, 204, 206, 207, 208, 210, 211, 212, 214, 215, 289, 354.
Handsome, 429.
Handwriting, 144.
Happiness, 473, 723.
Hardship, 213, 743, 769.
Haughty, 414.
Headache, 587, 633.
Health, 766.
Heart disease, 631, 676, 680.
Heraldry, 342.
Hereditary madness, 583.
Hindoos, 270.
Honour, 260, 770.
Hopeless passion, 613.
Horticulture, 803.
Hypochondria, 421.
Hypocrisy, 349, 556, 589.
Hysteria, 448, 502, 647.

I.
Idealism, 325, 326, 583, 708.
Ideality, 314, 315.
Idleness, 431.
Ill-health, 524, 606.
Illegitimacy, 613.
Ill-luck, 114, 426, 604, 718, 719.
Illness, 527, 530, 590.
Ill-success, 427.
Imprisonment, 602, 671.
Imprudence, 582, 758.
Impulse, 123, 189, 221.
Inconstancy, 175, 498, 511, 607, 625, 752, 791.
Indigestion, 598.
Infamy, 260.
Infidelity, 572.
Inheritance, 329, 790.
Innocence, 348.
Inquiry, 121.
Inquisitiveness, 136, 152.
Intelligence, 119, 266.
Intuition, 146.
Inventors, 124, 286.
Irony, 253.
Irresolution, 198.
Irritating subject, 215.

J.
Jack of all trades, 342, 620.
Jealous, 553.

318 INDEX.

Jews, 347.
Joints, 118, 119, 120, 121, 122, 130, 136, 147.
Journalist, 253.
Jupiter, Mount of, 405, 429; Charlatanry, 439; Doctors, 439; Handsome, 429; Marriage, 429; with other Mounts, 439.
Justice, 308.

K.
Knowledge, Search after, 240.

L.
Laborious life, 767.
Laplanders, 261.
Large thumbed subject, 200.
Lasciviousness, 644.
Laziness, 191, 247, 754.
Levity, 791.
Liar, 692.
Liberality, 246.
Libertine, 515.
Life, 545, 631, 667, 750.
Lines, 119, 126, 139, 404, 405, 410, 424, 425, 426, 523, 524, 526, 527, 529, 530, 531, 533, 534.
Literary critics, 515, 806.
Liver complaint, 555.
Logic, 182, 184, 242, 449.
Losses, 612, 702.
Love, 211, 516, 565, 610, 782.
Luxury, 146, 432.
Lying, 171, 771.

M.
Malignity, 256, 749.
Marriage, 240, 435, 452, 478, 518, 788; Happy, 435, 599, 656, 687; Unhappy, 533, 567, 663, 787.
Mars, Mount of, 398, 439, 475; Capacity for keeping temper, 477; Cowardliness, 483; Cruelty and violence, 480; Feeble line of heart, 482; Hot blooded, 477; Lines on mount, 480; Love of war, 481; Marriage, 478; Plain of Mars, 475; Sensual, 477; with Apollo, 485; with Mt. of Moon, 485.

Materialism, 146, 455.
Maternity, Dangers of, 568.
Mathematics, 231.
Meanness, 574.
Mechanics, Talent for, 243.
Meddlesomeness, 251.
Medium, 783.
Melancholy, 180, 414.
Melody, 807.
Mental agitation, 421.
Merchants, 122, 342.
Mercury, Mount of, 405, 429, 460; Athlete, Good, 462; Barristers, Good, 461; Charlatan, 465; Children, No. of, 469; Clairvoyant, 462; Commerce, 472; Contradiction, 474; Good fortune, 467; Happiness, 473; Hypochondria, 468; *Liaison*, 469; Mathematics, 463; with Apollo, 471; with Saturn, 474; with Venus, 473.
Minutiæ, Instinct for, 135.
Mischief, 141, 715, 760.
Misery, 114, 601.
Misfortune, 114, 701, 721, 785.
Mocking disposition, 149.
Moderation, 304.
Modesty, 715.
Moon, Mount of, 439, 486; Angle on, 505; Boundaries of, 497; Clairvoyance, 495; Crescent in woman, 505; Croix mystique, 495; Cross barred, 504; Drowning, 505; Gout, 498; Idleness, 496; with other Mounts, 506.
Moral force, 164.
Morbidness, 227.
Moslem tribes, 326.
Mount, Crosswise lines on, 404; Development of, 398; Displacement of, 407; Prevailing, 428; Two lines on, 403.
Murder, 593, 681.
Murderers, 194.
Murderous tendency, 570, 794.
Musicians, 121, 162, 273, 303.

Mystery of birth, 548, 683.

N.
Nails, 786, 794.
Narrow-mindedness, 112.
Nerves, 333.
Nervous temperament, 206, 254, 421, 581, 703, 751.
Nobleness, 740.
North American, 281.

O.
Observant, 136.
Obstinacy, 193, 691.
Occultism, 231, 310, 566, 634.
Old age, 630.
Orators, 236.
Orderliness, 119, 121, 122, 276, 301.
Organic affection, 415.
Orientals, 331.
Ostentatious, 430.

P.
Pain, 602.
Palm, 108, 112, 113, 114, 115.
Paradoxicalism, 119.
Passion, 292.
Perseverance, 282, 297.
Personal merit, 599, 623.
Phalanges, 146.
Philosophy, 158, 240, 318.
Physical attributes, 111.
Pleasure, Pursuit of, 430, 648, 798.
Poetry, 121, 203, 583, 589, 806.
Political freedom, 285.
Politics, 158, 285, 685.
Positivism, 235.
Poverty, 532, 625.
Practical, 157, 351.
Presentiments, 783.
Presumption, 173.
Pride, 221.
Promptness, 248.
Prophecy, Gift of, 327.
Prosperity in arms, 717.
Protestantism, 280.
Public entertainers, 429.
Pugnacity, 141.
Punctuality, 121, 302.

Q.
Quarrelsome disposition, 744.

INDEX. 319

Quickness, 132, 413.

R.
Reason, 119, 120, 121, 134, 146, 185, 285, 317, 322, 574.
Reflection, 121.
Religion, 166, 223, 224, 335, 627.
Republican, 218.
Resolution, 271.
Restlessness, 173, 701.
Riches, 239, 562, 618, 643, 734, 735, 746.
Roman Catholics, 280.
Romancists, 330.
Roué, 555.
Rudeness, 754.

S.
Sadness, 231, 441.
Sanguine, 412.
Saturn, Mount of, 405, 433, 440, 442; Aggressiveness, 449; Asceticism, 441; Bony fingers, 442; Cross on, 435; Happy, 448; Hysteria, 448; Immodesty, 449; Individualizing, 448; Logic, 449; Marriage, 435; Remorse, 441; Single straight line, 444; Spot, 445; Worry, 446; with other Mounts, 448, 449.
Sceptics, 317.
Science, 121, 203, 244, 245.
Sculpture, 805.
Second sight, 636.
Secretiveness, 246.
Sedentary, 287.
Seduction, 789.
Self-centred, 440.
Self-control, 573.
Self-reliance, 202, 271, 582.
Selfishness, 113, 178, 347.
Sensitive, 154, 595.
Sensual, 113, 126, 146, 221, 314, 315, 477.
Sentiment, 123.
Short life, 557.
Sickness, 635.
Signs, Corroboration of, 419.

Singers, 162.
Skin, 442.
Social freedom, 319.
Soldiers, 549.
Somnambulist, 638.
Sophist, 589.
Sorrow, 602, 746, 776.
Soul, 120, 329.
Spiritualist, 331.
Spots, 415, 458, 436.
Sterility, 469, 547, 563, 720.
Strength of will, 571, 576, 694.
Stubbornness, 150.
Success, 433, 434, 584, 610, 614, 674, 723.
Sudden death, 529, 536, 557, 567, 576, 710.
Sudden heroism, 187.
Sudden wealth, 774.
Suicide, 231.
Superstition, 231, 696, 701.
Suspicious, 247.
Symmetry, 121.
Sympathy, 333.
Synthesis, 133, 142.

T.
Tact, Want of, 582.
Tassel, 423.
Taste, 126.
Tenderness, 507.
Thief, 730, 795.
Thumb, 116, 165, 174, 176, 182, 200, 265, 272.
Tidiness, 159, 251.
Timidity, 580.
Toothache, 592.
Tranquil life, 766.
Treachery, 257, 560.
Truth, 225, 316, 317, 762.
Tyrants, 194.

U.
Uncertainty, 272.
Unchastity, 434, 715, 728.
Ungovernable passion, 194.
Unintelligence, 214.
Unreasonableness, 193, 559.
Unremitting labour, 558.
Unsuccessful, 564.
Uprightness, 215, 726.
Useful, 297.

Uselessness, 569.
Utopian ideas, 325.

V.
Vanity, 248, 598, 771.
Venus, Mount of, 405, 439, 507; Absence of Mount, 512; Asthma, 521; Beauty, Feminine form of, 507; Gallantry, 507; Good fortune, 516; Happiness, 520; Liberality, 520; Libertine, 515; Love, 516; Marriage, 518; Tenderness, 507.
Versatility, 569.
Violent death, 539, 601 (see Death).
Virago, 252.
Vivacity, 632.
Voyages, 773.

W.
Want of logic, 189.
Want of principle, 249.
Weak constitution, 727.
Weak digestion, 633.
Weak head, 759.
Weak heart, 577.
Weak intellect, 169, 585.
Wealth, 235, 446, 614, 616, 658.
Well-regulated mind, 119, 121, 306.
Widowhood, 784.
Will, 182, 184.
Woman, 784, 798; Affectionate, 363; Coquettish, 358; Curious, 362; English, 361; Fussy, 364; Haters, 560; Idealistic, 368; Influence of, 661; Knotty fingers in, 355, 356, 365; Love, 357; Pleasure loving, 360; Punctuality, 364; Sagacity, 357; Tyrannical, 365; Vivacious, 366.
Worry, 138, 504, 533, 622, 781.
Wounds, 405, 541, 565, 585, 591, 592, 596, 672, 673, 713.